# Digital Design

# Logic, Memory, Computers

## Electrical and Electronic Engineering Design Series

Books written by Nicholas L. Pappas, Ph.D.

Electric Circuits – Analysis and Design

Electronic Circuit Design – with Bipolar and MOS Transistors

CMOS Circuit Design – Analog, Digital, IC Layout

Digital Design – Logic, Memory, Computers

Analog Filter Design

## Mathematics

Books written by Nicholas L. Pappas, Ph.D.

Arithmetic – Integers, Fractions, Decimals

Algebra – A Clear Presentation

# Digital Design
## Logic, Memory, Computers

Nicholas L. Pappas, Ph.D.

**A Message about this Text:** The subject is essentially endless. The purpose here is to say enough about the subject so that you, the reader, have a running start when you apply this knowledge to your work.

Algebraic and Circuit Analysis skills are useful here.

We believe important benefits accrue by doing the problems carefully, by reconstructing the Truth Tables and K maps to and from functions, by reworking the text equations if only to copy them, by drawing the circuits, block and timing diagrams, and the very important effort doing the experiments. These efforts provide startup experience.

Once you have some work experience we are confident that you will be able to expand your know how with reasonable effort.

**A Message from the Author:** I have worked continuously in the electronics industry since 1950 except for 11 semesters teaching at San Jose State University (Professor and Chair Computer Engineering 1988-1993). There I discovered my talent for teaching such as it may be. After War2 I attended Lehigh University, and then transferred to Stanford where I earned the MS degree and, while working at HP in the early 1950's, the Ph.D. EE degree. (Somehow I did not get the word and formally apply for the BS degree.) Hardware design has been my principal activity. I learned enough about assembly language, Forth, C and C++ to design the software I needed for my projects. My current activity is designing integrated circuits.

# Preface

This text is different from the many introductory digital design texts, because we actually design a product by *implementing a design* and not *just talk* about logic circuits used in a digital circuit. And, we ask you to work hard doing experiments so that you acquire real world experience with commercially available digital circuits. In other words this is about real learning.

To be fair, the contents of this text are only a beginning, because modern digital circuits are assemblies of *thousands of logic gates* where the traditional schematic is essentially useless. Consequently traditional gate level circuit design, the schematic capture method, of a thousand gate circuit is impractical. The solution to this design problem is a *hardware description language* or HDL. The HDL language we prefer is Verilog. However there is a catch.

> You have to know how to design using schematic capture before you can design using Verilog's text capture method. Design using schematic capture is what this text is about. We provide a start showing how to design with Verilog's text capture method so that you can move on to multi 1,000 gate chip designs.

We start at the beginning by presenting a top down design method. Our concern here is the design of *digital systems*. Newer names for digital systems are Tablet, and Smart Phone.

All digital systems are *programmed* to perform tasks. The *programs* may simply be represented by push button switches on a front panel. At the other extreme is a general purpose computer whose programs are a set of *digital words* stored in a *memory* that define one or more tasks.

We jump into the water by starting immediately with a top down design of *Euclid's gcd Calculator*. The *Euclid's gcd Calculator* design process is a vehicle for learning in this text. The process provides context for topics as they are presented thereby avoiding the question what is that for? The design proceeds step by step to produce the product design as we acquire the know-how we need.

Digital Design

The design method divides and conquers by recognizing that any digital system consists of two major parts – *the controller* and *the controlled* known as the *control and datapath* shown here as black boxes (Figure 1).

**Figure 1 The control/datapath structure of a digital system**

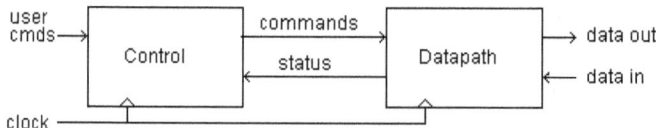

We learn about three basic tools necessary to execute any digital design. Two tools are the Truth Tables and Karnaugh maps, which are graphical displays of switching functions. The third tool is the theoretical basis of digital design - a Switching Algebra.

The basic circuits of digital logic are building blocks *without* memory. They are standard commercially available logic circuits, which are described and their equations are presented. The *experiments* include applications of these building blocks.

The ASM (algorithmic state machine) chart, a fourth tool, is a preferred way to represent the algorithm. We show how to implement Algorithmic State machine (ASM) charts, how to synchronize an input, how to link ASMs, and how to make state assignments.

We move forward evolving the design of *Euclid's gcd Calculator* by creating and implementing more complete ASM charts for the product.

The design of complex building blocks *with* memory is based on elementary blocks with one bit memory known as flip-flops. The designs are implemented by the ASM method. We show how to design up and up/down synchronous counters, shift registers, and linear feedback shift registers, LFSRs. And, the design of *Euclid's gcd Calculator* continues.

This is followed by showing how to design memory systems with and without a cache hierarchy. ASM charts and associated timing diagrams allow us to readily implement the designs. These are charts and timing diagrams we have not found anywhere else.

A high reliability memory system incorporates error correction and control (ecc) circuits. We show how to add ecc circuits to a memory system, discuss the required mathematics, and describe the coding process.

Maurice Wilkes invented what is now called microprogramming about 1950. Wilkes' ingenious idea simplified state machine circuit designs by implementing the state machine circuit as a matrix of diodes. As time elapsed this 1950's format was converted to *modern microprogramming* format in a straightforward way.

A computer has two basic parts – cpu control and a datapath for processing instructions. We define a user instruction set (uI), the uI address modes, and how the uI are formatted as binary words. Status bits and their condition codes that implement program control are defined. We show how each uI is represented by a list of micro instructions mI that is executed by the datapath, and how the datapath executes the mI under cpu control.

Verilog is a large language with many parts. We only discuss the basic parts. For large circuit designs text capture of digital designs is preferred to schematic capture. Verilog uses text capture to represent digital circuits with a hierarchy of modules that are interconnected via input and output ports. We show by example how to write modules defining digital circuits so that you can move on to multi 1,000 gate chip designs using Verilog, which is classified as a hardware description language (HDL).

Chapter abstracts are next.

> Our blog *npappasee.blogspot.com* may offer you additional information. Take a look.

> We would appreciate receiving your comments and views on this text at *npappasz@yahoo.com.*

Digital Design

*A Top Down Design Method*  The design method is described by using an example (a Euclid gcd Calculator). The method divides and conquers by first recognizing that any digital system consists of two major parts – *the controller* and *the controlled* known as the *control and datapath*

A description is written and then is converted into a specification that is transformed into an *algorithm*. An example of an algorithm, Euclid's gcd algorithm, is executed by hand so that the process is clear. In turn the algorithm is transformed into an *Algorithmic State Machine (ASM)* chart, A front panel for the Euclid gcd Calculator is created so that we know what we want as a final product. The ASM chart is readily converted into a design of the control and datapath. As the process proceeds documentation is produced.

Note: this design is more efficiently implemented using a microprocessor, but that is not the point here.

**1 *Truth Tables*** All digital systems and circuits are some combination of the digital gates AND, OR, and NOT. A digital circuit has inputs and outputs whose relationships are represented by a mathematical equation referred to as a switching function such as    $F(x, y, z) = xy' + y'z' + x'yz$ .
A truth table is a graphical display of a switching function. A truth table has a finite size, because logic variables, such as x, y, z, can take only two values 0 and 1.

We start with the truth tables for AND, OR, and NOT digital gates. As an example we include the widely used XOR gate that is an assembly of AND, OR, and NOT gates.

We show how to convert a switching function equation to a truth table, and conversely, a truth table to a switching function equation. Truth table size grows rapidly: 5 input variables produce 32 truth table lines, 6 variables produce 64 lines and so forth. We show how to reduce truth table size by using the important idea of *Table Entered Variables*.

An important issue is that any design specification is unavoidably incomplete. The graphical form of the *truth table* facilitates visualizing the specification. One advantage of truth tables is that incompleteness is expressed as missing output entries. This is convenient, because you fill in these entries by making logical deductions after you acquire some experience. Furthermore, you can experiment with different 0,1 entries.

**2 *Karnaugh Maps*** A *Karnaugh map* (K map) is a graphical display of a switching function that presents the same information shown in a truth table. The map is an array of squares. A function's input variables are the K map's coordinates. The function's output is shown as 0 or 1 in each square. We believe a K map provides the easiest way to simplify switching equations. The useful K map *cluster* idea facilitates circuit simplification and conversion to equations. We show how to convert a switching function to a K Map, and how to convert a K Map to a switching function. K maps grow rapidly in size. We show how to reduce K map size by using the important idea of *Map Entered Variables*. Incomplete word descriptions or specifications result in missing entries in Karnaugh map squares. With experience you can provide the missing 0 or 1 entries with reasonable estimates so that the design process can proceed. The missing entries are referred to as *don't care* terms.

**3 *Switching Algebra*** Digital circuits process signals which are in one of two states, high H or low L, at any instant of time. A mathematical method facilitates analysis and design of digital circuits. The mathematical method presented here is a Switching Algebra of two valued variables such as 0 and 1. As in Euclid's geometry the switching algebra is based on a set of axioms. The axioms are used to prove theorems. Then axioms and newly proven theorems are used to prove more theorems, and so forth. Then the ideas of *minterms* and *maxterms* representing function terms are presented. The chapter closes with a discussion of *Perfect Induction*, switching function truth tables, and the number of switching functions.

**4 *Building Blocks with no memory*** We do not waste your time with non-standard building blocks. We use real circuits available from industry. Building blocks with no memory implement logic equations as OR of ANDs that is a sum of *minterms* (e.g. $m_1+m_3+m_7$) or AND of ORs that is a product of *maxterms* (e.g. $M_1M_3M_7$). A *no-memory* building block's distinctive characteristic is that output signal values are computed from *present* input signal values. A change of any input value produces new outputs after a *time delay* required to compute new values. This *propagation delay* depends upon the implementing technology.

Digital signals are either at H(igh) voltage or L(ow) voltage. We show how assigning 1 and 0, in different *mixed logic* combinations, to the H and L states of nodes defines 8 functions each for NAND, NOR, and XOR gates. Magic? No. Just logic. Mixed logic recognizes that a node voltage can be assigned true when H *or* L in a real circuit. Mixed logic simplifies circuit analysis and design.

A variety of building blocks are discussed (see Table of Contents). A black box with inputs and outputs is shown for each standard block. And, *defining function equations* and related timing diagrams are presented. Any circuit is designed without difficulty from its equations.

Boolean equations represent *steady state* behavior. One might think that Boolean equations imply correct transient response to changes in inputs. In fact Boolean equations per se do not guarantee correct transient response. We show how *hazards and glitches* arise, and what to do about them by using *timing diagrams*. Timing diagrams are graphics illustrating the behavior of a circuit's logic as a function of time.

**5 *State Machines*** A state machine implements an algorithm. The ASM chart (algorithmic state machine chart) is a preferred way to represent the algorithm. The ASM chart is drawn using lines and three symbols: the *state rectangle*, the *input diamond*, and the *conditional output oval*. The modern algorithmic state machine design procedure converts the ASM chart into a logic circuit. The method is straightforward. No tricks or black magic are required. Conversely a circuit can be converted to an ASM chart. The difficult creative part of the design is producing the specification that describes the function(s) to be implemented.

Inputs asynchronous to a clocked state machine introduce the possibility of metastability. Input signal synchronization is illustrated.

The idea of linked state machines simplifies complex ASM charts and their implementation. A linked state machine implements a function *once* in a design. A function that can be used over and over again.

Clare's method provides a reasonable solution to the important problems of *state assignment* and *races*.

The Euclid gcd algorithms are designed. They illustrate the many details that arise when designing a functioning state machine circuit.

**6 *One Bit Building Blocks with Memory*** When a *synchronous state machine* clock changes state, new stored values result after a delay. (A clock is just another input.) Then new values appear at the outputs after another delay required computing them from present inputs and the new stored values.

Memory is implemented in logic circuits by *feedback* of a circuit output to a circuit input. The RS Latch is the most elementary one-bit circuit with memory. Latches do not have clock inputs.

The ideas of *Setup and Hold Windows* guarantee correct *synchronous* circuit operation.

Edge Triggered flip-flops are one-bit memory circuits with clock inputs. They are *synchronous* circuits.

The D and JK positive edge triggered flip-flops are the standard industry flip-flop circuits. The T flip-flop is derived from the JK flip-flop.

Flip-flop truth tables are derived from the flip-flop defining equations.

*A flip-flop's next state defining equation is the most important fact a designer needs to know.*

We discuss why metastability can be the Achilles' heel of digital systems.

**7 *Complex Building Blocks with Memory*** These are state machines consisting of a register and logic gates. A *register* is a group of n flip-flops storing n bits that relate to each other in some way as defined by the wiring of the logic gates. The ASM method simplifies analysis and design of building blocks with memory such as counters, shift registers, and linear feedback shift registers.

Counting registers report (store) the number of events received. Standard counting building blocks have up to four counter control lines: clear to zero, load data, count events, count up or down. An ASM chart defines priorities and control line actions. We show how to use a flip-flop's *next state defining equation* to design a 3 bit up counter, and a 3 bit up/down counter in standard building block format. We stay with 3 bits so that a focus on the ideas is not lost in a flood of too many bits. Other counter applications illustrate design of *no-glitch digital waveform generators*.

A *shift register* has the property that all bits stored in the register can move, on the same clock edge, one position to the left (or right) of their present position by asserting control lines. We design shift registers by applying the same methods used to design counters with emphasis on the *next state defining equation and ASM charts.*.

A Linear Feedback Shift Register (LFSR) is an n-bit counter that has a *pseudo-random* state sequence. An LFSR is a shift-register circuit that includes one or more xor gates. LFSR design requires knowledge of Galois field theory that we introduce and point the way to further study. We do this to illustrate how sophisticated mathematics can be applied to digital circuit design. There is no need to hack the LFSR design process. And, the design of *Euclid's gcd Calculator* concludes.

**8 *Memory*** Every useful digital circuit is operated by a state machine, and every state machine uses memory. The memory may be just a one bit register, or $2^{25}$ bits supporting a computing state machine. One bit, or some group of n bits, represents a *word* of data stored in the memory. Memory stores words for use on demand.

A memory *data path,* considered as a black box, is connected to an address bus, an i/o (input/output) bus, and control lines from a state machine. This is the minimum set of connections. The memory may have more than one i/o port. The memory may consist of just one memory chip or be a complex assembly of chips including memory chips, address decoders, buffers, and so forth. Memory devices implement a large number of registers. One disadvantage is that only one register at a time (one memory word at a time) can be accessed in a memory chip.

The primary memory parameter is the *access time* that is the time required to execute a read or write operation. System timing diagrams illustrate the importance of access time to computer system performance.

The design of a memory system is presented: the system block diagram, the ASM chart, the timing diagrams, the controlling finite state machine, the address decoders, the address and data registers, the array of memory chips.

*One level of complexity* Memory system cost reductions have led to creation of *memory hierarchy designs: cache backed up by main memory.*

Cache memory systems and design are implemented: the ASM Chart, the timing diagrams, the Cache Data Path Architecture, and the Cache Control.

*Other levels of complexity*    A high reliability system incorporates error correction and control (ecc) circuits. R. W. Hamming's ingenious idea initiated the development of ecc circuit design.

Code design is a very complex subject using very complex mathematics.

We show why parity bits are required to create code words. We discuss block codes, explain Hamming's ingenious idea, and show how the BCH codes expanded Hamming's codes to be able to correct more than one error.

A code design process is presented, and means to implement it is described.

**9 *Microprogrammed State Machines*** Maurice Wilkes invented micro-programming about 1950. He showed how any state machine can be physically realized by what was later called *microprogramming*. He recognized that logic design naturally separates into two parts: the control state machine, and the controlled data path. Wilkes' ingenious idea simplified state machine circuit designs by converting them into a matrix of diodes format. Later this 1950's format was converted to *modern microprogramming* format in a straightforward way.

**10 *CISC-8*** The goal is the design a *complex instruction set computer* (CISC). We show how that is done. We start by defining a *complex user instruction set* (uI), the address modes for reading and writing memory, and the status bits for branch decisions. The bit by bit format of each uI is specified so that the computer will execute the correct actions for that uI.

Each user instruction is implemented as a *list* of micro-instructions (mI), because a uI is too complex to execute in one step when systematic circuits are used (this is not obvious). Executing a mI list executes the uI it represents as a series of elementary steps.

We show how to create the user datapath and datapath control from the user instruction set.

**11** *Verilog - a Language for Digital Design* Now that gate level design of some elementary and some complex algorithms has been presented in Chapters 1 to 10 we turn to the major design issue of the day. *How do we design a digital system when the implementation requires a very, very large number of gates?*

For large circuit designs *text capture* of digital designs is preferred, and superior, to *schematic capture*. The reasonably straightforward Verilog *text capture* syntax offers clarity and savings in design time.

The intent is to show how standard digital circuits are implemented by Verilog code. Programming experience is required to follow what is presented. This chapter is only an introduction to Verilog.

*Fourteen Experiments*
The Solderless Breadboard and the Power Supply

Experiment 1 NAND, NOR, NOT, XOR Gates
Experiment 2 Multiplexers
Experiment 3 Decoders
Experiment 4 Binary Arithmetic
Experiment 5 Comparing Numbers
Experiment 6 Flip-flops
Experiment 7 Counters
Experiment 8 Shift Registers
Experiment 9 Single Pulse, Debounced, Synchronized
Experiment 10 System Clock
Experiment 11 Receiver Serial Clock
Experiment 12 SRAM Read/Write ASM to circuit
Experiment 13 Cache ASM to circuit
Experiment 14 Hamming Code Encoder and Decoder

RMS Voltage

# Contents

# The SI System of Units
## SI Prefixes

| Prefix | Multiplier | Symbol |
|--------|-----------|--------|
| Tera | $10^{12}$ | T |
| Giga | $10^{9}$ | G |
| mega | $10^{6}$ | M |
| kilo | $10^{3}$ | K |
| milli | $10^{-3}$ | m |
| micro | $10^{-6}$ | μ |
| nano | $10^{-9}$ | n |
| pico | $10^{-12}$ | p |
| femto | $10^{-15}$ | f |
| atto | $10^{-18}$ | a |

## Basic Units

| Quantity | Name | Symbol |
|----------|------|--------|
| Length | meter | m |
| Mass | kilogram | kg |
| Time | second | s |
| Current | ampere | A |
| Temperature | Kelvin | K |
| Luminous Intensity | candela | cd |

## Derived Units

| Quantity | Name | Formula | Symbol |
|----------|------|---------|--------|
| Acceleration | meter per sec per sec | $m/s^2$ | a |
| Velocity | meter per sec | $m/s$ | v |
| Force | newton | $kg{\times}m/s$ | N |
| Pressure (stress) | pascal | $N/m^2$ | Pa |
| Density | kg per cubic meter | $kg/m^3$ | r |
| Energy or work | joule | $N{\times}m$ | J |
| Power | watt | $J/s$ | W |
| Charge | coulomb | $A{\times}s$ | C |
| Potential | volt | $W/A = J/C$ | V |
| Resistance | ohm | $V/A$ | Ω |
| Capacitance | farad | $C/V$ | F |
| Magnetic flux | weber | $V{\times}s$ | Wb |
| Inductance | henry | $Wb/A$ | H |

# Greek Alphabet

| | | | |
|---|---|---|---|
| A | α | alpha | a[1] |
| B | β | beta | b |
| Γ | γ | gamma | g |
| Δ | δ | delta | d |
| E | ε | epsilon | e |
| Z | ζ | zeta | z |
| H | η | eta | h |
| Θ | θ | theta | q |
| I | ι | iota | i |
| K | κ | kappa | k |
| Λ | λ | lambda | l |
| M | μ | mu | m |
| N | ν | nu | n |
| Ξ | ξ | xsi | x |
| O | o | omicron | o |
| Π | π | pi | p |
| P | ρ | rho | r |
| Σ | σ | sigma | s |
| T | τ | tau | t |
| Y | υ | upsilon | u |
| Φ | φ | phi | f |
| X | χ | chi | c |
| Ψ | ψ | psi | y |
| Ω | ω | omega | w |

---

[1] equivalent computer keyboard English letter keys

# A Top Down Design Method

Our concern here is the design of *digital systems*, digital machines, or whatever name you like. Digital systems have been designed with names Traffic Controller, Timer, Black Jack Game, Counter, Printer, Computer, Tablet, and Smart Phone.

Where to begin? Experience informs us to begin at the top in order to avoid the large and small traps of beginning elsewhere.

All digital systems are *programmed* to perform tasks. The *programs* may simply be represented by push button activated switches on a front panel of a black box, which when pressed execute their assigned tasks. At the other extreme is a general purpose computer whose programs are a set of *digital words* stored in a *memory* that define one or more tasks. Tasks activated by complex action(s) initiated by pressing keyboard keys.

Fortunately there is a straightforward general method that allows you to design *any* digital system you may be asked to design. The World calls it A Top Down Method. Here is how we implement the method.

The method divides and conquers by first recognizing that any digital system consists of two major parts – *the controller* and *the controlled* known as the *control and datapath* shown here as black boxes (Figure 1).

**Figure 1 The control/datapath structure of a digital system**

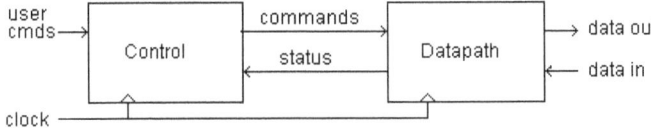

A user commands the control to execute a specific action. The control issues commands to the datapath. The commands 'setup' the datapath to perform operations on data at the end of the clock period. Then the datapath sends to the control the status of the results of the operation. This cycle of events continues until a stop command occurs automatically or is issued by the user.

Digital Design

*Data out* are inputs to devices the digital system was actually designed to operate in the first place (such as print a page, machine a metal part). *Data in* is feedback from the devices.

**1 *The Design Process*** A project is started by creating a description of the system to be created. A description is then converted into a specification that is transformed into an *algorithm*. In turn the algorithm is transformed into an *Algorithmic State Machine (ASM)* chart, which is implemented by digital electronic hardware. As the process proceeds documentation is produced.

> *An algorithm is a procedure, requiring no creative skills of the user, with precise instructions, specifying a finite number of steps, so that sooner or later the procedure ends.*

**2 *Example of an Algorithm*** We will show how to implement Euclid's algorithm as a digital machine after we learn how to execute it by hand. *Euclid's algorithm* provides you with the means to find the greatest common divisor (gcd) of any two integers m and n. The method is based on the fact that the last step of a sequence of division steps produces a zero remainder (Euclid proved that this always happens). Note that the *next step is executing division of the **reciprocal** of the remainder term of the present step (m/n is replaced by n/r).* We execute Euclid's algorithm on the number pair 585, 165. *The last divisor* 15, with zero remainder, is the gcd.

$$\frac{m}{n} = q + \frac{r}{n} \quad \text{where } q \text{ is the quotient and } r \text{ is the remainder}$$

$$\frac{m}{n} = q_1 + \frac{r_1}{n} \qquad \frac{585}{165} = 3 + \frac{90}{165} \qquad 585 = 165 \cdot 3 + 90$$

$$\frac{n}{r_1} = q_2 + \frac{r_2}{r_1} \qquad \frac{165}{90} = 1 + \frac{75}{90} \qquad 165 = 90 \cdot 1 + 75$$

$$\frac{r_1}{r_2} = q_3 + \frac{r_3}{r_2} \qquad \frac{90}{75} = 1 + \frac{15}{75} \qquad 90 = 75 \cdot 1 + 15$$

$$\frac{r_2}{r_3} = q_4 + \frac{r_4}{r_3} \qquad \frac{75}{15} = 5 + \frac{0}{15} \qquad 75 = 15 \cdot 5 + 0$$

The gcd of 585 and 165 is 15. Divide both numbers by the gcd to form the reduced fraction. (Exercise: find the gcd of 100 and 27)

$$\frac{585}{15} = 39 \quad and \quad \frac{165}{15} = 11 \quad \Rightarrow \quad \frac{585}{165} = \frac{15 \times 39}{15 \times 11} = \frac{39}{11}$$

Here is a general statement of Euclid's algorithm. Given two positive integers M and N, find their greatest common divisor.

1. Find the remainder. Divide M by N to get remainder R ($0 \leq R < N$).
2. Test remainder R for zero. If R=0 end the process and N is the gcd. If R≠0 go to step 3.
3. Replace M with N, and N with R. Go to step 1.

**3 *Algorithmic State Machine*** Next we implement Euclid's algorithm as an ASM chart (Figure 2), which we explain here and on pages 4, 5.

First know that the chart moves from state x to state y at the *end* of each clock period. The clock is *not* shown on the ASM chart.

**Figure 2 Euclid ASM**

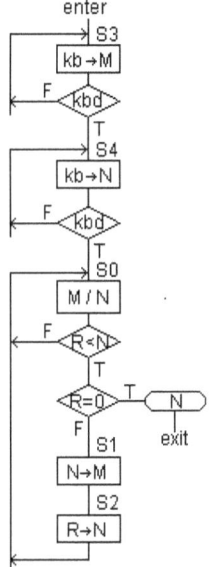

Enter the ASM chart at state $S_3$. Use the keyboard to store a number in M. Go to state $S_4$ when *kbd is T*. Use the keyboard to store a number in N. Go to state $S_0$ when *kbd is T*.

Divide M by N while in state $S_0$ to produce R. The ASM remains in $S_0$ until the division process completes when (*R<N*) becomes True. Decision diamonds *R<N* and *R=0* are activated while the ASM is in state $S_0$. Decisions execute at the end of each clock period.

*Simultaneously* while in state $S_0$ test *R<N* for T or F. If *R<N* is False go to state $S_0$.

*Simultaneously* while in state $S_0$ test remainder R for zero. If *R=0* AND *R<N* are True output the value of N and exit the ASM. If *R<N* is True AND *R=0* is False go to state $S_1$ in the next clock period.

While in state $S_1$ replace M with N. Go to $S_2$ in the next clock period.

While in state $S_2$ replace N with R. Go to state $S_0$ in the next clock period.

*kb→M, kb→N, M/N, N→M*, and *R→N* are commands sent to the datapath, while *kbd, R=0* and *R<N* are status signals sent to control (Figure 1).

## Algorithmic State Machine (ASM) Charts

The Algorithmic State Machine[2] (ASM) chart represents a synchronous or an asynchronous finite state machine (FSM). The ASM chart (Figure 2 for example) is drawn using lines and three symbols: the *state rectangle*, the *input diamond*, and the *conditional output oval*. A synchronous machine's clock is not shown on the ASM chart in order to simplify the chart. States are represented by rectangles (e.g. $S_0$ in Figure 2). The rectangle has only one entrance and one exit. Any number of paths may lead to a state rectangle's entrance. However, only one path leads away from a state rectangle's exit. The path goes directly to another state when this is an unconditional change of state as is $S_1$ to $S_2$. Or, the path enters a logic tree of conditions (one or more diamonds such as $R=0$ and $R<N$). Each condition has a corresponding logic circuit output (from the diamonds) to its next state.

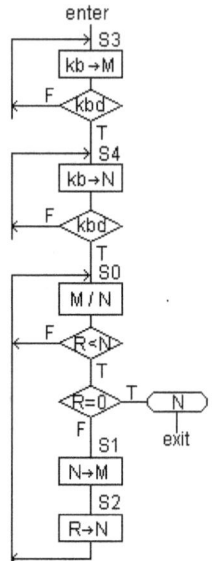

**Figure 2 Euclid ASM**

**State variable** State variables are the outputs of the state register. For example (binary bit) outputs $q_2$, $q_1$ and $q_0$ are state variables.

**State** The state of a machine is the binary number representing the current combination of bits stored in the state register such as $q_2q_1q_0$.

**State assignment** The state assignment process assigns a different number to each state to identify each state uniquely. The set of numbers is the state assignment. This is an important step, because the assignments affect the complexity of the next state circuits and output circuits.

**Unconditional Branch** If no input variables (no diamonds) are associated with the present state (such as $S_1$ in Figure 2), then the next state the machine enters is determined only by the present state ($S_1 \rightarrow S_2$).

[2] Winkel, D., F. Prosser. *The Art of Digital Design*. ISBN 0130467804
Clare, C. R. *Designing Logic Systems Using State Machines*,
ISBN 0701111200

*Synchronous Machine Clock* (not shown in the ASM chart). The clock action is as follows. At every rising clock edge the next state decision is made to step from the present state to the next state. The present state and the inputs' values (T or F) at the clock edges determine which state is the next state. For example, in Figure 2 when the present state is $S_0$, if ASM input $R<N$ is False the next state is $S_0$. Follow the path from $S_0$'s exit through the $R<N$ diamond's F-exit back to $S_0$'s entrance.

*Diamonds* represent branch decisions. An *external* input is entered inside a diamond such as *kbd, R<N,* and *R=0* in Figure 2. The diamond symbol has True and False exit paths that implement a branch decision. Instead of one variable (branch) equation(s) of several variables can be placed in the diamond. Inputs to a diamond may be inputs from the outside world, or they may originate from within the same state machine.

*Conditional Branch* If one or more input variables (in diamonds) are associated with the present (current) state, then the next state the machine goes to depends on this present state, and the present values of the input variables. Input variables not associated with this present state have a don't care status when the state machine is in this present state.

*Outputs* Any state can be assigned no outputs, one or more unconditional outputs (in state rectangles), and, or one or more conditional outputs (in ovals). The same output can be unconditional in one state and conditional in another state.

*Unconditional outputs* are listed in state rectangles. Unconditional outputs do not depend on inputs in any way. Unconditional outputs are asserted for every clock period the associated state is in the present state. For example, in Figure 2, $M/N$ is asserted for every clock period the ASM is in state $S_0$.

*Conditional outputs* are shown in ovals that are always associated with a branch diamond representing the condition. Conditional outputs are asserted for every clock period the path from the present state, via one or more diamonds, to the oval entrance is asserted. For example, in Figure 2, the N conditional output is asserted *during* the present state when the present state is $S_0$, and the expression $R<N$ *AND* $R=0$ is True.

**4** *Euclid's gcd Calculator* Given Euclid's gcd algorithm we can conjure up a front panel of the black box containing the calculator. A keyboard is used to enter the numbers M and N (Figure 3). We press push button M to direct key presses of digits representing M to the M display and to the M memory. We enter number N the same way. Then press gcd to calculate the gcd. Press M to start again.

**Figure 3 The Euclid gcd Calculator Front Panel**

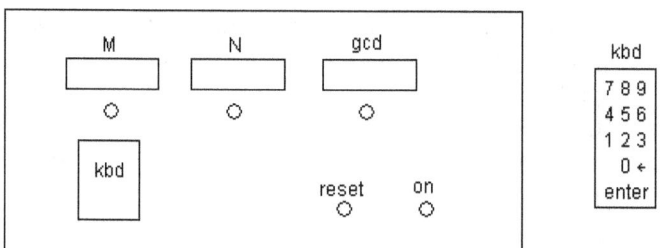

*The creative, and difficult, part of the design is producing the specification, and the ASM, that describes the function(s) to be implemented. Implementation of the ASM as a synchronous state machine is a straightforward process.*

**5** *Digital Electronic Hardware* Digital systems are assembled from the many commercially available digital building blocks. This is why you buy and use them. You do not have design them. Designing digital building blocks is a totally different activity.[3]

Boolean algebra is the mathematical basis of digital design, because it is an algebra of base 2 numbers 0 and 1. Compare binary base 2 numbers to the decimal base 10 numbers 0 to 9. Base 2 is used, because 2 level, on/off, circuits are physically realized without difficulty.

In a physical system the voltage levels H and L (High and Low) at circuit nodes correspond to the Boolean logic values 1 and 0 (true and false) or vice versa, or both (at different nodes). Base 2 digits are referred to as binary digits 0 and 1, which are also called bits.

We now proceed to show that the ASM chart contains the information that allows you to design the *datapath* and the *control*. Given the ASM chart you can actually design the entire system with every detail accounted for.

---

[3] Nicholas L. Pappas, *CMOS Circuit Design – Analog, Digital, IC Layout*

**6 *Designing the Control*** From the ASM we derive and use a *truth table*, which is a graphical display of the logic equations describing a digital circuit. The truth table has columns for ASM states, decision variables, ASM next states, unconditional state outputs, and conditional outputs (Table 101). We will derive the control equations from the truth table.

The 3 state variables $q_2q_1q_0$ are represented by 3-tuples (1,0 strings) $000_2$, $001_2$, $010_2$, $011_2$, $100_2$, $101_2$, $110_2$, $111_2$, which are base 2 binary numbers.

The decimal equivalents are 0, 1, 2, 3, 4, 5, 6, 7. We use the + symbol to indicate the next state as $q_2^+q_1^+q_0^+$. In this circuit only $S_0$ to $S_4$ are used.

An output is true (T or 1) or false (F or 0) in a state for each combination of associated input values. Input values are represented by 0 or 1. For example, there are four possible combinations of values $00_2$, $01_2$, $10_2$, $11_2$, when there are two inputs such as $R<N$ and $R=0$.

**Table 101 Euclid's ASM (dash – is a don't care , output 0's omitted)**

| Present State $q_2q_1q_0$ | kbd | R<N | R=0 | Next State $q_2^+q_1^+q_0^+$ | N | kb $\rightarrow$ M | kb $\rightarrow$ N | M / N | N $\rightarrow$ M | R $\rightarrow$ N |
|---|---|---|---|---|---|---|---|---|---|---|
| 000 | – | 0 | – | 000 | | | | 1 | | |
| 000 | – | 0 | – | 000 | | | | 1 | | |
| 000 | – | 1 | 0 | 001 | | | | 1 | | |
| 000 | – | 1 | 1 | – (exit) | 1 | | | 1 | | |
| 001 | – | – | – | 010 | | | | | 1 | |
| 010 | – | – | – | 000 | | | | | | 1 |
| 011 | 0 | – | – | 011 | | 1 | | | | |
| 011 | 1 | – | – | 100 | | 1 | | | | |
| 100 | 0 | – | – | 100 | | | 1 | | | |
| 100 | 1 | – | – | 000 | | | 1 | | | |

In order to design the control circuit, equations for the next state number are required. Before we do that we need to know how L and H voltage levels relate to 1, 0 true and false logic levels.

---

*Problem* 1 What is the logic equation that keeps the ASM in State 3?

*Problem* 2 What is the logic equation that produces output N?

---

**7 *A digression into Mixed Logic*** In Figure 4 a bubble at a node means that node signal is true (1) when the node voltage is low (L). Absence of a bubble at a node means that node signal is true (1) when the node voltage is high (H). H and L voltages representing logic true at different nodes is referred to as mixed logic. As will become clear mixed logic produces simplified logic circuits that are easier to understand.

If we assign 1 to a node when it is H, then L must be 0 at that same node.

A *not* gate creates active high and active low nodes (Figure 4d, 4e). Here active means true. E.g. if the source of *kbd* is H we use circuit e to get an active low *kbd*.

In Figure 4a when both *and* inputs are H the output is H. If either, or both, *and* inputs are L the output is L. The wires from the S3 and kbd nodes to the gate inputs are HH wires (*Wires and Logic* page 9).

In Figure 4b when both *and* inputs are L the output is H. If either, or both, *and* inputs are H the output is L. The wires from the S3 and kbd nodes to the gate inputs are LL wires

In Figure 4c when both *and* inputs are L the output is H. The wire from the S4 node to the gate input is an LL wire. This next one is different. The wire from the kbd node to the gate input is an HL wire. The HL wire implements the *not* function.

**Figure 4 Positive and negative True Logic**

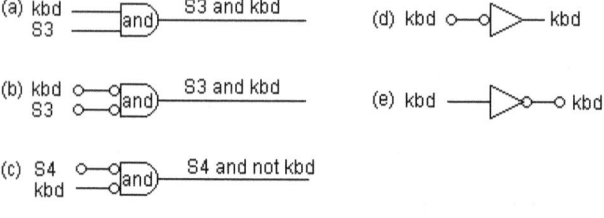

---

*Problem* 3 Write the truth table for the *and* circuit.

*Problem* 4 Write the truth table for the *not* circuit.

*Problem* 5 Write the truth table for the *or* circuit.

---

### Wires and Logic

We intend to show here that logic circuit analysis is facilitated if the two ends of a wire are considered to be two independent nodes, when one node is a gate output and the other node is a gate input. This point of view allows us to compare active logic levels at the two nodes of a wire.

A wire from the active low inverter output z-bubble to a $gate_2$ input connects an active low output node to an active high input node. We call this an LH wire. Input z to $gate_2$ is active low, whereas input $a_0$ is active high. The $a_0$ wire is an HH wire. Two other HH wires are the wires from active high z and $a_1$ connected to $gate_1$ active high inputs. Observations such as this gave rise to the idea to consider two ends of a wire as two independent nodes.

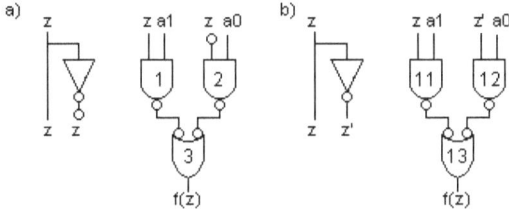

The output of $gate_2$ goes L when *both* inputs z and $a_0$ are H. If $a_0$ is H, then z must be H to drive the $gate_2$ output to L. In other words, z must be false in order to be H.

To show a z false equivalent, we remove the bubble marking z as active low, and *replace z bubble with z tick (z')* as shown at $gate_{12}$.

$gate_2$ *out* = $gate_{12}$ *out* = $z'a_0$

The LL wires from $gate_1$ and $gate_2$ outputs to the $gate_3$ inputs do not create ticks. HH and LL wires do not create ticks (complements).

---

*conclusion*
Tick variables on LH or HL wires. Do not tick variables on LL or HH wires

---

**8 *Designing the Control continued*** Each line of Truth Table 101 adds terms to the equations for the control. It turns out that, since each line is in some state, state numbers $q_2q_1q_0$ will appear in the logic equations for the control. It will be convenient to use a 3 input to 8 output decoder, because it produces one output line for each state number $q_2q_1q_0$ (Figure 5).

**Figure 5 Decoder**

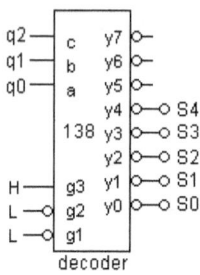

The 3 digit binary state number is stored in 3 one bit circuits for which we select D flip-flops (page 12). One next state equation is required for each flip-flop (Figures 6, 7, 8). Table 101 is the source of the equations. We tick variables that are logical 0 in an equation. For example tick kbd as *kbd'*. Tick is the logical *not* symbol. The logical *and* symbol is ×. The logical *or* symbol is +. In each $q^+$ column of Table 101 there *happen to be* two 1's. This means each equation has 2 terms. Usually they are different numbers.

($q_2^+$) When ($S_3$ *and* kbd) OR ($S_4$ *and* kbd') is True, then $q_2$ is set to 1 on the next clock edge. The equation is $d_2 = S_3 \times kbd + S_4 \times kbd'$ (Figure 6).

**Figure 6 Gate Circuit for $q_2$ flip-flop**

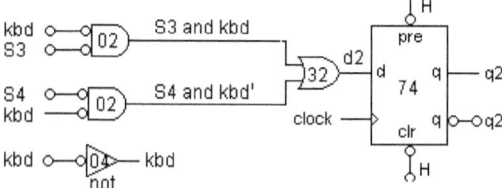

($q_1^+$) When ($S_1$) OR ($S_3$ *and* kbd') is True, then $q_1$ is set to 1 on the next clock edge (Figure 7). $d_1 = S_1 + S_3 \times kbd'$

**Figure 7 Gate Circuit for $q_1$ flip-flop**

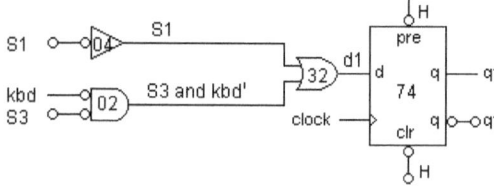

($q_0^+$) When ($S_3$ *and* kbd') OR ($S_0$ *and* R<N *and* R=0') is True, then $q_0$ is set to 1 on the next clock edge (Fig 8). $d_0 = S_0 \times (R < N) \times (R = 0)' + S_3 \cdot kbd'$

**Figure 8 Gate Circuit for $q_0$ flip-flop**

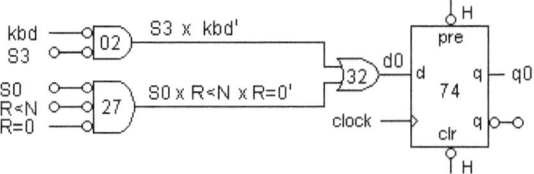

Output equations are also derived from Table 101 with one term per 1 bit.

$$M/N = s_0 \qquad N \to M = S_1 \qquad R \to N = S_2$$

$$kb \to M = S_3 \qquad kb \to N = S_4 \qquad N = S_0 \times (R < N) \times (R = 0)$$

Combining all figures and equations we get Figure 9. *However this is not a product yet, because the M and N pushbuttons are NOT shown here. To be continued.*

**Figure 9 Euclid Control including ASM Outputs**

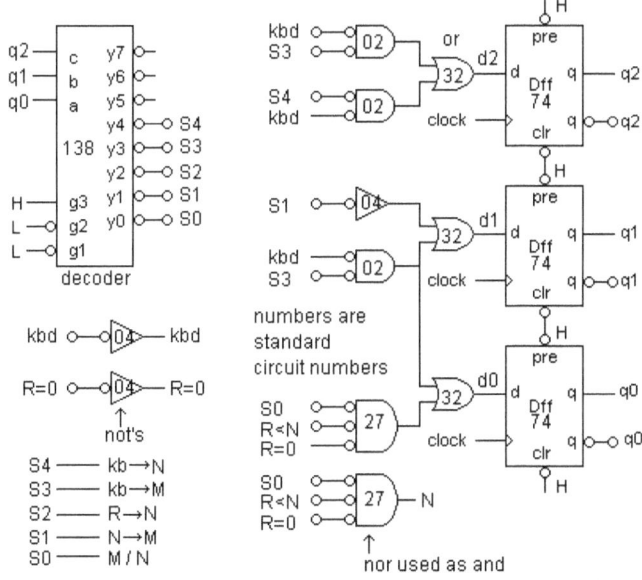

---

*Problem* 6 Find a commercially available digital circuit for each circuit in Figure 9.

---

*Clock* A clock is an independent signal source that emits a periodic signal.

In a digital system the clock wave-form is essentially a squarewave (Figure 504a). The transition from L to H is referred to as the positive edge (marked as up arrows in Figure 504c), and the transition from H to L is referred to as the negative edge (marked as down arrows in Figure 504d). The transition times are referred to as the rise and fall times respectively. Idealized waveforms with zero transition times make timing diagrams more readily perceptible (Figure 504b).

The time between two consecutive positive, or two consecutive negative, transitions is the clock's *period T* that is the reciprocal of the clock's *frequency f*. A 500MHz clock has a 2ns period (T=1/f). We emphasize that there is one

**Figure 504 Periodic Clock Waveform**

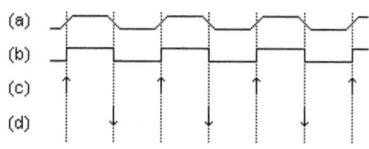

*triggering edge* per period by showing the clock waveform as a series of up arrows (Figure 504c), *or* down arrows (Figure 504d).

## D positive edge triggered flip-flop

The D flip-flop ($D_{FF}$) building block has a synchronous input clock, and data input d. The d input defines the next state (Figures 506, 507). Asserting the asynchronous preset or clear inputs sets or resets the q outputs, while overriding the synchronous inputs. While the asynchronous preset or clear inputs are not asserted, the defining equation $f_d=d$ is operational. The (present state) data at the d input meeting setup and hold requirements are transferred to the (next state) output q on the next positive clock edge (Figure 507). Triggering occurs when the edge rises up past a voltage level, and is not dependent on edge's rise time over a wide range. After the hold time interval elapses data d may be changed with no effect on the outputs.

The $D_{FF}$ has two output lines for q. One output is active high and the other is active low. The truth table for the $D_{FF}$ includes inputs pre, clr, d, and the clock. The up arrow symbol means the defining equation is executed on the positive edge of the clock.

| $D_{FF}$ | Inputs | | | | Outputs | |
|---|---|---|---|---|---|---|
| pre | clr | clk | d | q | q' |
| L | L | – | – | * | * |
| L | H | – | – | H | L |
| H | L | – | – | L | H |
| H | H | ↑ | L | L | H |
| H | H | ↑ | H | H | L |
| H | H | L | – | $q_o$ | $q'_o$ |

**Figure 506 Positive edge triggered D flip-flop**

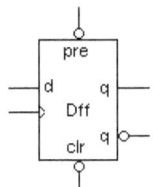

*both H but unstable,

$q_o$ is present state, $q$ is next state $q^+$

The dashes mean don't care. The $D_{FF}$ defining equation is $\boxed{f_d = d}$

**D flip-flop timing diagram** If any clock period n is the current period then it is the *present period*, and clock period n+1 is the next period. The value of d in the present period is the *present state of d*. Since q⁺=d the output in clock period n+1 is the *next state of q* that is a copy of present state d (Figure 507). Thus the q waveform is the same as the d waveform delayed in time by one clock period. (A one clock period slip assumes the d input is synchronous with the clock.) We do not show a q⁺ waveform, because in the next state q⁺ becomes the present state q.

**Figure 507 D flip-flop timing diagram**

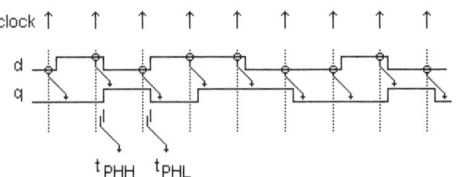

The coincidence of the d values and the clock edge are marked by circles (Figure 507). The circles mark the cause and arrows show the effect. The propagation delays from clock to output must be greater than zero. When $t_{PHL}$ and $t_{PLH}$ (Figure 507) are not greater than zero, then the d and q waveform transitions merge with the clock edge. Then equation q⁺=d cannot answer the question What is the value of d at the clock edge? When propagation delays are zero there is no definite answer. This is why propagation delays greater than zero are necessary for synchronous operation. Since all physical circuits have propagation delays greater than zero this is not a problem.

**9 *Designing the Datapath*** The Euclid ASM informs us of the control commands and the datapath status reports (Figure 10). The control commands setup the datapath to execute the commands. As we proceed we will discover additional commands and status reports simply because we cannot predict or think of everything at once.

**Figure 10 The control/datapath structure of the Euclid ASM**

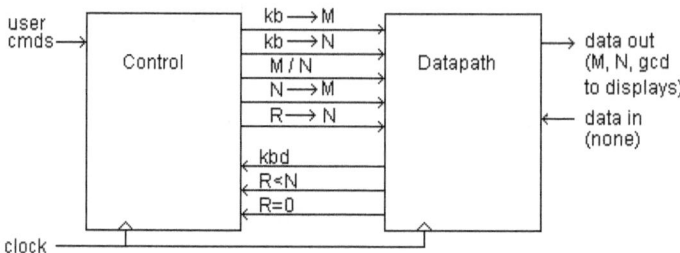

We analyze and think about the ASM details, with the front panel in mind, in order to determine what should be added to the datapath.

We need the keyboard for M and N data entry. Keyboards are designed to send out a *binary* number when a key is pressed. For example type the digits 6, 7, 8, 9 to send the 6, 7, 8, 9 sequence of binary numbers to the z bus via *sbz* (Figure 11 page 15). Use backspace ($\leftarrow$) to change digits. Press Enter to terminate the entry process.

The black box kbX implements the *kb→M* and *kb→N* commands.

Pressing Enter also orders the kbX black box to convert the digits 6, 7, 8, 9 into the single number 6,789 in binary format, and store it as a base 2 number in the M memory. Repeat the conversion for N. In this way numbers M and N are stored in the datapath. Stored where?

Numbers M, N, and R have to be available over several clock periods, while the algorithm is executing. Therefore numbers have to be stored in a *memory*, because a memory carries data from one clock period to the next. A *register* is a one-number-memory. Registers are required for M, N, R and gcd. Displays are required for M, N, & gcd. Add a zero detector for R.

## Figure 11 ASM Datapath major components

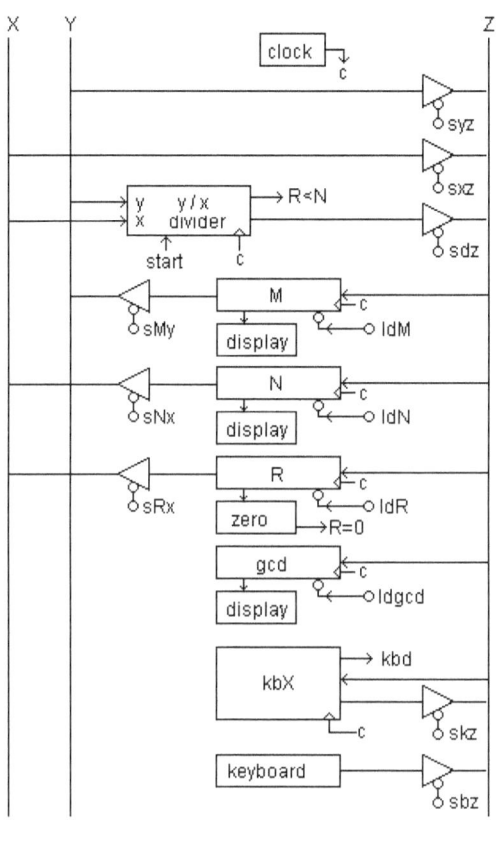

M/N requires a number *divider y/x.*

Data is transferred on buses *x, y, z.* On/off *tri-state gates allow only one number at a time to be on a bus.* E.g. *N→M* is implemented when *control* sends the commands *sNx, sxz, ldM* to the *datapath.* This set of commands loads N into M.

(1) *sNx, sxz* set up a path from register N outputs to the z bus and M inputs.

(2) in the next clock period *ldM* loads N into the M register. (the z bus is the input to all registers).

Here we have a block diagram of the database where each line, such as x, y, z, as drawn is actually n wires that make up a *bus.* There is one wire for each bit.

---

*Problem* 7 What is the set of datapath inputs that replace M by N?

*Problem* 8 What is the set of datapath inputs that replace N by R?

*Problem* 9 What is the set of datapath inputs that divide M by N and store the result in R?

---

At this point, *in order to focus on the ideas*, we have omitted various items, such as the pushbuttons, from the datapath. Furthermore we have avoided references to specific datapath hardware until we study the basic standard commercial circuits. Consider what has been presented so far as an introduction to the design process.

In *5.6 Euclid's gcd Calculator* we will develop two out of three parts of what we hope is a finished product ASM. Hope becomes reality only after we build and successfully test the product. The third part is developed in 7.5.

The ASM has three parts. The first part takes in the digits for M and N from the keyboard, stores them as digits, and displays them so that you know what you have entered. The second part converts the digits into *numbers* M and N. Finally the gcd of M and N is calculated and displayed.

The conversion process is a mathematical operation we present to show how the conversion is actually done. First in general terms, and then by specific example of the greatest 4 digit number 9999. Knowing this we can then design for the datapath a digital circuit implementing the process.

Then we do the same for the actual gcd calculation. The specific example uses the numbers 585 and 165 from the example done by hand.

# 1 Truth Tables

All digital systems and circuits are some combination of the digital gates AND, OR, NOT. A digital circuit has inputs and outputs, which are related by mathematical equations referred to as switching functions such as $F(x, y, z) = xy' + y'z' + x'yz$ .

*A truth table is a graphical display of a switching function.* A truth table has a finite size, because logic variables, such as x, y, z, can only have the values 0 and 1.

We show how to convert a switching function equation to a truth table, and conversely, a truth table to a switching function equation.

Truth table size grows rapidly: 5 input variables produce 32 truth table lines, 6 variables produce 64 lines and so forth. We show how to reduce truth table size by using the important idea of *Table Entered Variables*.

An important issue is that any design specification is unavoidably incomplete. One advantage of truth tables is that incompleteness is expressed as missing output entries. This is convenient, because you fill in these entries with 0 or 1 by making logical deductions after you acquire some experience. Furthermore, you can experiment with different 0,1 entries.

A truth table has N input columns for N variables, $2^N$ rows, and one output column for each function F. The input columns are filled in by listing in numerical order the $2^N$ binary numbers 0 to $2^N-1$. If N=3, then 3 input columns are filled in with 8 numbers $000_2$ to $111_2$ in 8 rows ($0_{10}$ to $7_{10}$). Subscript 2 means the number is a base 2 number. The function F is evaluated for each combination of variables' values, and the 0 or 1 results are entered into the output column.

The number of input and output variables is estimated from the design specification that also dictates how the output columns are filled in. This process requires experience. We begin by converting switching equations to truth tables and vice-versa after presenting the logical operators AND, OR, NOT, XOR.

# 1.1 Truth Tables for AND, OR, NOT, and XOR

The operator symbols × or · (dot), +, and ' (tick) used in the axioms (Chapter 3) are also known as the operator words AND, OR, and NOT respectively. Operator symbols × or · (multiply), + (plus), and ' (tick) replace operator words.

$f(x, y) = x \ AND \ y = x \cdot y = x \times y = xy \quad (can \ omit \cdot or \times)$

$f(x, y) = x \ OR \ y = x + y$

$f(y) = NOT \ y = y'$

***Two level voltage system*** Gate circuits are energized by a DC power supply. Gate circuit terminals are at 0 and 1.8 or 0 and 5 volts, or some other pair of voltage levels that depend on the implementing technology. Gate circuit outputs switch between high and low levels referred to as H and L. In gate circuits the two levels may not reach the power supply levels, because high level and low level device source impedances prevent this when current flows. Actual H and L levels represent ranges of voltages (Figure 101). The ranges depend upon the technology used. Circuit node H and L levels must not be in the forbidden region. The forbidden region separation provides a noise margin that can prevent errors when noise interferes.

**Figure 101**

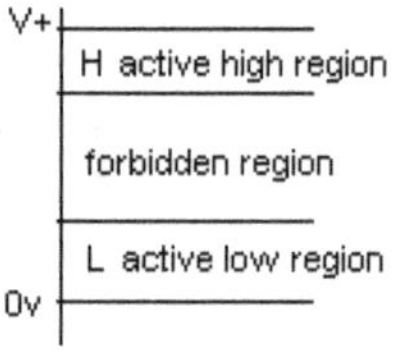

In a physical system the voltage levels H and L at nodes correspond to the Boolean logic values 1 and 0 (true and false) or vice versa, or both (at different nodes). Binary digits 0 and 1 are referred to as bits.

***Truth Tables for 2 input Gates*** A gate output is true (1) or false (0) for each combination of input variable values. Input variable values are represented by 0 or 1. For example, there are four possible combinations of values when there are two input variables x, y. The combinations of values are written as 1,0 strings also known as n-tuples. Two inputs x, y have 2-tuples $00_2$, $01_2$, $10_2$, $11_2$ as combinations of values interpreted as base 2 binary numbers whose decimal equivalents are 0, 1, 2, 3.

Upcoming is a truth table with two input columns x, y and *three* output columns xy, x+y, y'.

The 0, 1 format is logical and the H, L format is physical. Here are the AND, OR, NOT truth tables.

| Logical | | | AND | OR | NOT | Physical | | | AND | OR | NOT |
|---|---|---|---|---|---|---|---|---|---|---|---|
| Row | $x$ | $y$ | $xy$ | $x+y$ | $y'$ | Row | $x$ | $y$ | $xy$ | $x+y$ | $y'$ |
| 0 | 0 | 0 | 0 | 0 | 1 | 0 | L | L | L | L | H |
| 1 | 0 | 1 | 0 | 1 | 0 | 1 | L | H | L | H | L |
| 2 | 1 | 0 | 0 | 1 | | 2 | H | L | L | H | |
| 3 | 1 | 1 | 1 | 1 | | 3 | H | H | H | H | |

The *row* numbers in a truth table are the decimal equivalents of the binary numbers representing the 0, 1 assignments to input variables on each row. Decimal row numbers are more useful than one might think. Suppose there are 64 rows generated by six input variables. For most of us recognizing $53_{10}$ is easier than recognizing $110101_2$.

How are the values in the AND, OR, NOT output columns determined? One can show formally how they are determined (Chapter 3 Switching Algebra). An informal, common sense method might proceed as follows.

Assign H = True = 1, and L = False = 0.
*AND*
If x is true and y is true, then AND is true.      $(f=xy=HH=1\times1=1)$
If x is true and y is false, then AND is false      $(f=xy=HL=1\times0=0)$
If x is false and y is true, then AND is false      $(f=xy=LH=0\times1=0)$
If x is false and y is false, then AND is false      $(f=xy=LL=0\times0=0)$

*OR*
If x is true or y is true, then OR is true.      $(f=x+y=H+H=1+1=1)$
If x is true or y is false, then OR is true      $(f=x+y=H+L=1+0=1)$
If x is false or y is true, then OR is true      $(f=x+y=L+H=0+1=1)$
If x is false or y is false, then OR is false      $(f=x+y=L+L=0+0=0)$
This is the inclusive OR where any true input makes the output true.

*NOT*
If x is true, then NOT is false      $(f=x'=H'=L=0)$.
If x is false, then NOT is true      $(f=x'=L'=H=1)$.

# Digital Design

**Schematics** Gate schematics are constructed from a small set of parts: wires (lines), bubbles (circles), triangles, AND shapes, and OR shapes.

**Figure 102 Basic parts of electronic gate symbols**

We use basic parts to create schematics for AND, OR, and NOT gates (Figures 102, 103). (AND and OR gates may have more than two inputs.)

**Figure 103 AND, OR, NOT gates**

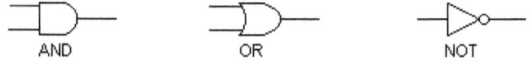

**Exclusive OR** The important Exclusive OR function XOR is derived from AND, OR, and NOT. The 2 input XOR is an example of a composite operator requiring one OR, two AND, and two NOT operators (Figure 104). The XOR output is true when all inputs are not equal.

A 2 input XOR is described as follows. Output f is true when x is false AND y is true, OR, when x is true AND y is false. I.e. when x and y are not equal the XOR output is true. The 2 input XOR *defining equation* is

$$f(x,y) = x'\,y + xy' = x \; xor \; y = x \oplus y \quad (f \; equals \; x \; XOR \; y)$$

| Row | x | y | x' | y' | x'y+xy' | $f(x,y)$ |   | x | y | $x \oplus y$ |
|-----|---|---|----|----|---------|----------|---|---|---|------|
| 0 | 0 | 0 | 1 | 1 | 1·0+0·1 | 0 |   | 0 | 0 | 0 |
| 1 | 0 | 1 | 1 | 0 | 1·1+0·0 | 1 | $\Rightarrow$ | 0 | 1 | 1 |
| 2 | 1 | 0 | 0 | 1 | 0·0+1·1 | 1 |   | 1 | 0 | 1 |
| 3 | 1 | 1 | 0 | 0 | 0·1+1·0 | 0 |   | 1 | 1 | 0 |

A straightforward gate implementation of the 2 input XOR defining equation is shown in Figure 104. There are other XOR implementations that use fewer parts.

**Figure 104 XOR gate**

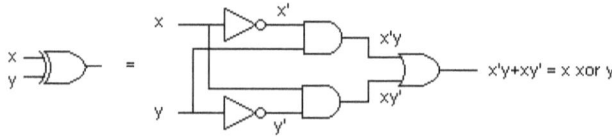

20

## 1.2 Equation to Truth Table

Converting a known function, such as F(x,y,z), to a truth table requires calculating the 0 or 1 value of the function F for each combination of input variables. Three input variables have $2^3 = 8$ combinations of values $000_2$ to $111_2$. Consequently the truth table has 8 rows, 3 columns for input variables x, y, z, and one column for output F. Converting the equation $F(x, y, z) = xy' + y'z' + x'yz$ to a truth table requires evaluating F eight times, once for each row.

In row 0, where x, y, & z equal 0, $F(x, y, z) = xy' + y'z' + x'yz$ is evaluated as $F(0,0,0) = (0 \times 1) + (1 \times 1) + (1 \times 0 \times 0) = 1$. Here is the table with F' (not-F) added.

| Row | x | y | z | $xy' + y'z' + x'yz$ | F | F' |
|-----|---|---|---|---------------------|---|----|
| 0 | 0 | 0 | 0 | $0 \cdot 1 + 1 \cdot 1 + 1 \cdot 0 \cdot 0$ | 1 | 0 |
| 1 | 0 | 0 | 1 | $0 \cdot 1 + 1 \cdot 0 + 1 \cdot 0 \cdot 1$ | 0 | 1 |
| 2 | 0 | 1 | 0 | $0 \cdot 0 + 0 \cdot 1 + 1 \cdot 1 \cdot 0$ | 0 | 1 |
| 3 | 0 | 1 | 1 | $0 \cdot 0 + 0 \cdot 0 + 1 \cdot 1 \cdot 1$ | 1 | 0 |
| 4 | 1 | 0 | 0 | $1 \cdot 1 + 1 \cdot 1 + 0 \cdot 0 \cdot 0$ | 1 | 0 |
| 5 | 1 | 0 | 1 | $1 \cdot 1 + 1 \cdot 0 + 0 \cdot 0 \cdot 1$ | 1 | 0 |
| 6 | 1 | 1 | 0 | $1 \cdot 0 + 0 \cdot 1 + 0 \cdot 1 \cdot 0$ | 0 | 1 |
| 7 | 1 | 1 | 1 | $1 \cdot 0 + 0 \cdot 0 + 0 \cdot 1 \cdot 1$ | 0 | 1 |

*Problems* 101-108 Make truth tables for the following functions.

101 f = xy'

102 f = x + y'

103 f = x ⊕ y'

104 f = (x + y)(x + y')

105 f = x + xy

106 f = x(x + y)

107 f = (x + y)(x' + z')(y + z)

108 f = xy + x'z + yz

## 1.3 Truth Table to Equation

We show how to derive F(x,y,z) from this truth table.

| Row | x | y | z | F | Four ways from table to equation |
|-----|---|---|---|---|----------------------------------|
| 0 | 0 | 0 | 0 | 1 | *1 outputs sum of products* |
| 1 | 0 | 0 | 1 | 0 | $f = m_0 + m_3 + m_4 + m_5$ |
| 2 | 0 | 1 | 0 | 0 | *0 outputs sum of products* |
| 3 | 0 | 1 | 1 | 1 | $f' = m_1 + m_2 + m_6 + m_7$ |
| 4 | 1 | 0 | 0 | 1 | *1 outputs products of sums* |
| 5 | 1 | 0 | 1 | 1 | $f' = M_0 M_3 M_4 M_5$ |
| 6 | 1 | 1 | 0 | 0 | *0 outputs products of sums* |
| 7 | 1 | 1 | 1 | 0 | $f = M_1 M_2 M_6 M_7$ |

We leap ahead here. Each row represents a *minterm m* (e.g. xy'z) or a *maxterm M* (e.g. x'+y+z') (Section 3.3.1 page 51, Section 3.3.2 page 54).

Minterms and maxterms have one occurrence of *every* function variable, or its complement in any term. The row number becomes the row's minterm or maxterm number (e.g. for row 5 the symbols are $m_5$, $M_5$).

Minterms are true for only one combination of variables (such as xy'z), because a minterm is an AND function. On the other hand maxterms are true for all combinations of variables except one, because a maxterm is an OR function. I.e. a maxterm is false for only one combination of variables (such as x'+y+z'). Know that $m_5$ = xy'z is the *logical complement* of $M_5$ = x'+y+z' (Theorem 13d page 45).

In upcoming paragraphs the equation for F is written as a sum of minterms representing the 1's of F, or a product of maxterms representing the 0's of F.

Function F(x,y,z) above is true for truth table rows 0,3,4,5. Clearly F is true (F=1) when one or more minterms $m_0$, $m_3$, $m_4$, $m_5$ are true. F is false (F=0) only when *all* of the four minterms are false. This is why F=$m_0$+$m_3$+$m_4$+$m_5$ is a *sum* of minterms (logical OR).

Function F(x,y,z) above is false for truth table rows 1,2,6,7. Therefore F is false when *any* of the maxterms $M_1$, $M_2$, $M_6$, $M_7$ are false (any M=0).

Conversely F is true only when *all four* maxterms are true (all M=1). This is why $F=M_1M_2M_6M_7$ is a *product* of maxterms (logical AND).

> Important: the *sequence* of variables in the following must be x, y, z.

We start the conversion process by *focusing on the 1's* to derive the *minterm canonical form* after offering a suggestion for minimizing errors when writing out minterms.

*Suggestion: Consider minterm $m_5$*

| | |
|---|---|
| 1. *Write down the variables:* | $xyz$ |
| 2. *Enter the minterm number:* | $101$ |
| 3. *Tick the zeros:* | $xy'z$ |

$F = m_0 + m_3 + m_4 + m_5$  $\rightarrow$  *$f = 1$ in rows $0,3,4,5$*

1. $F = xyz + xyz + xyz + xyz$

2.     $000 + 011 + 100 + 101$

3. $F = x'y'z' + x'yz + xy'z' + xy'z$

Simplify (this requires experience)

$F = m_0 + m_4 + m_3 + m_4 + m_5$

$F = x'y'z' + xy'z' + x'yz + xy'z' + xy'z$

$F = (x' + x)y'z' + x'yz + xy'(z' + z)$     (*now use $1 = x + x'$*)

$F = y'z' + x'yz + xy'$

Next we *focus on the zeros* to derive the *maxterm canonical form*. First, here is a suggestion for minimizing errors when writing out maxterms.

*Suggestion: Consider maxterm $M_5$*

| | |
|---|---|
| 1. *Write down the variables:* | $x + y + z$ |
| 2. *Enter the maxterm number:* | $101$ |
| 3. *Tick the ones:* | $x' + y + z'$ |

$F = M_1M_2M_6M_7$     $\rightarrow$  *$f = 0$ in rows $1,2,6,7$*

1. $F = (x + y + z)(x + y + z)(x + y + z)(x + y + z)$

2.     $001$        $010$        $110$        $111$

3. $F = (x + y + z')(x + y' + z)(x' + y' + z)(x' + y' + z')$

Digital Design

The following is an example of the complexities of simplification. Numbers following equations are theorem numbers (Chapter 3 page 45). We prefer using Karnaugh maps (Chapter 2).

$$F = (x + y + z')(x + y' + z) \times (x' + y' + z)(x' + y' + z')$$
$$F = (x + y + z')(x + y' + z) \times (x' + y') \qquad 10d$$

*Multiply out and simplify:*
$$F = (x + y + z')(x + y' + z)(x' + y')$$
$$F = [(x + y)(x + y') + (x + y)z + z'(x + y') + z'z](x' + y') \qquad \textit{multiply out}$$
$$F = [(x + xz + yz + z'x + z'y' + 0](x' + y') \qquad 10d$$
$$F = [x + x(z + z') + yz + z'y'](x' + y')$$

$$F = [x + yz + y'z'](x' + y')$$
$$F = xx' + x'yz + x'y'z' + xy' + yzy' + y'z'y'$$
$$F = 0 + x'yz + x'y'z' + xy' + 0 + y'z'$$
$$F = x'yz + (x' + 1)y'z' + xy'$$
$$F = x'yz + y'z' + xy' \qquad 7$$

*Emphasis: Four equations can be derived from any truth table* The equations are the answers to the following questions.

When does $f = 1$?  When does $f = 0$?
When does $f' = 1$?  When does $f' = 0$?

---

*Problems* 109-116 Make truth tables for the following functions ($g=1$).
109  *mux*  $f(z) = g(za_1 + z'a_0)$
110  *mux*  $f(y, z) = g(yza_3 + yz'a_2 + y'za_1 + y'z'a_0)$
112  *mux*  $f(x, y, z) = g(xyza_7 + xyz'a_6 + xy'za_5 + xy'z'a_4$
      $+ x'yza_3 + x'yz'a_2 + x'y'za_1 + x'y'z'a_0)$
113 Make one decoder truth table with 4 output functions.
      $y_0 = gy'z' \quad y_1 = gy'z \quad y_2 = gyz' \quad y_3 = gyz$
114 Make one decoder truth table with 8 output functions.
      $y_0 = gx'y'z' \quad y_1 = gx'y'z \quad y_2 = gx'yz' \quad y_3 = gx'yz$
      $y_4 = gxy'z' \quad y_5 = gxy'z \quad y_6 = gxyz' \quad y_7 = gxyz$
115 *Half adder* $s = x \oplus y$ *and* $c = xy$
116 *Full adder* $s = x \oplus y \oplus c_{in} \quad c = xy + c_{in}(x + y)$

---

## 1.4 Table entered Variables

The number of rows in a truth table doubles each time an input variable is added. The number of rows ($2^n$) is 2, 4, and 8 for n = 1, 2, and 3 input variables. Conversely the number of rows is halved when an input variable is converted to a *table entered variable*. Here is how you make the conversion.

**Basic idea** The coefficient of a term such as xyz in an algebraic expression is an implied 1. In any algebra any part of a term multiplied by the remaining part can be defined as a coefficient of the remaining part. For the term xyz, the coefficient of xyz is 1, the coefficient of yz is x, the coefficient of y is xz, and so forth.

**Example t xor q** A four row table is reduced to a two row table when input variable t in the $f = t\ xor\ q$ equation is converted to a table entered variable.

$f = t\ xor\ q$

| Row | t | q | f |
|-----|---|---|---|
| 0 | 0 | 0 | 0 |
| 1 | 0 | 1 | 1 |
| 2 | 1 | 0 | 1 |
| 3 | 1 | 1 | 0 |

*To convert t to a table entered variable:*

$f = m_1 + m_2 = t'q + tq'$

When $q = 0$: $f = t$  so enter $t$ in the $q = 0$ row

When $q = 1$: $f = t'$  so enter $t'$ in the $q = 1$ row

$\Rightarrow$

| Row | q | f |
|-----|---|---|
| 0 | 0 | t |
| 1 | 1 | t' |

verify : $f = tq' + t'q$

Other ways to do this is start with the equation. Write out the equation as a sum of minterms, and evaluate the equation directly in a truth table.

*If $f = t\ xor\ q$ then*

$f = t'q + tq' = t'(q) + t(q')$

$\Rightarrow$

| q | $t'q + tq'$ | f |
|---|-------------|---|
| 0 | $t' \cdot 0 + t \cdot 1$ | t |
| 1 | $t' \cdot 1 + t \cdot 0$ | t' |

Digital Design

***Example jq'+k'q*** A function of three variables has eight minterms. The eight row table is reduced to a four row table when input variable q is converted to a table entered variable. The trick is to form the four minterms of $f_{jk}$, and factor out coefficients q, q', 1, and 0. Note how axiom x+x'=1 is used to expand $f$.

$$f_{jk} = jq' + k'q$$
$$= jq'(1) + k'q(1)$$
$$= q'j(k + k') + qk'(j + j') \qquad axiom \ 4$$
$$= q'(jk) + (q' + q)(jk') + q(j'k')$$
$$= (q')jk + (1)jk' + (0)j'k + (q)j'k' \qquad [adding \ row \ 1 \ minterm \ m_1 = (0)j'k]$$

The coefficients are the $f_{jk}$ output values.

| Row | j | k | q | $f_{jk}$ |
|-----|---|---|---|----------|
| 0 | 0 | 0 | 0 | 0 |
| 1 | 0 | 0 | 1 | 1 |
| 2 | 0 | 1 | 0 | 0 |
| 3 | 0 | 1 | 1 | 0 |
| 4 | 1 | 0 | 0 | 1 |
| 5 | 1 | 0 | 1 | 1 |
| 6 | 1 | 1 | 0 | 1 |
| 7 | 1 | 1 | 1 | 0 |

$\Rightarrow$

| Row | j | k | $f_{jk}$ |
|-----|---|---|----------|
| 0 | 0 | 0 | q |
| 1 | 0 | 1 | 0 |
| 2 | 1 | 0 | 1 |
| 3 | 1 | 1 | q' |

In the 8 row truth table take two rows at a time. In rows 0, 1 f=q. In rows 2, 3 f=0. In rows 4, 5 f=1. In rows 6, 7 f=q'.

---

***Problems*** 117-119 Make truth tables with table entered variables (g=1).

117 *mux* $f(z) = g(za_1 + z'a_0)$

118 *mux* $f(y, z) = g(yz a_3 + yz'a_2 + y'za_1 + y'z'a_0)$

119 *mux* $f(x, y, z) = g(xyz a_7 + xyz'a_6 + xy'za_5 + xy'z'a_4$
$+ x'yz a_3 + x'yz'a_2 + x'y'za_1 + x'y'z'a_0)$

---

# Review 1

***AND, OR, NOT Operators*** The arithmetic operators $\times$ or $\cdot$, $+$, and $'$ are known as logical operators AND, OR, and NOT respectively. The XOR is derived from AND, OR, and NOT. The $\times$ or $\cdot$ may be omitted

| Row | $x$ | $y$ | AND $xy$ | OR $x+y$ | NOT $y'$ | XOR $x'y+xy'$ |
|-----|-----|-----|----------|----------|----------|----------------|
| 0 | 0 | 0 | 0 | 0 | 1 | 0 |
| 1 | 0 | 1 | 0 | 1 | 0 | 1 |
| 2 | 1 | 0 | 0 | 1 | | 1 |
| 3 | 1 | 1 | 1 | 1 | | 0 |

***Equation to Truth Table*** Every function has a truth table. The truth table has an input side listing all 0,1 combinations of input variables, and an output side listing a function's 0 or 1 value for each input combination. The truth table is constructed by (a) writing out $2^n$ rows of binary numbers in the range 0 to $2^n-1$ to representing combinations of the n input variables. (b) Evaluating the function's switching equation for each row to get 0 or 1 for each combination of input variables, and (c) entering the 0 or 1 in the same row of the output column representing the function.

***Truth Table to Equation*** Each row represents a minterm or a maxterm. The equation for the function F can be written as a sum of minterms or a product of maxterms. The switching equation is derived from the truth table by including minterms corresponding to rows for which F = 1 in the table. Or, the switching equation is derived from the truth table by including maxterms corresponding to rows for which F = 0 in the table.

***Table entered Variables*** The number of rows in a truth table doubles each time an input variable is added. Conversely the number of rows is halved when an input variable is converted to a table entered variable. The basic idea behind table entered variables is this. The coefficient of a term such as xyz in an algebraic expression is an implied 1. In any algebra any part of a term multiplied by the remaining part can be defined as a coefficient of the remaining part. For the term xyz, the coefficient of yz is x, the coefficient of y is xz, and so forth. *If $f = t \ xor \ q$ then $f = t'q + tq'$*

And, coefficients $t'$ and $t$ of q and $q'$ can be table entered variables.

# 2 Karnaugh Maps

A Karnaugh[1] map (K map) is a graphical display of a switching function (e.g. Figures 201, 202, 203, 204). The map is an array of squares. The function's input variables are the array's coordinates. The function's output is marked as 0 or 1 in the squares. We believe a K map provides the easiest way to simplify switching functions.

> *Each square of a K map is equivalent to a truth table row. In other words each square represents a minterm or maxterm. The squares are numbered so that square k corresponds to truth table row k.*

The switching function is represented by 0's and 1's in the squares: each square is assigned the same 0 or 1 output value of the switching function that is assigned to the corresponding row in a truth table.

The K map is a truth table alternative. As is the case with truth tables, incomplete word descriptions or specifications result in missing entries in Karnaugh maps. With experience you can provide the missing entries with reasonable estimates so that the design process can proceed.

## 2.1 A Map is a Graphical Display

K maps for functions of one, two, three, and four variables are in Figures 201, 202, 203, and 204. Various ways of labeling K map coordinates are shown, as well as base 10 and base 2 square numbers. Each way has its advantages that are discovered with experience.

In Section 2.7 (page 40) we show a method, *map entered variables*, of working with more than four variables. Processing of a very large number of input variables is facilitated by methods such as Quine-McClusky[2].

---

[1] Karnaugh, M. *The Map Method for Synthesis of Combinational Logic Circuits* Transactions A.I.E.E., Communications and Electronics 72, Part 1 (November 1953): 593-599.
[2] Quine, W. V. *The problem of Simplifying Functions American Mathematical Monthly* 59 (October 1952): 521-531.
Quine, W. V. *A Way to Simplify Truth Functions American Mathematical Monthly* 62 (November 1955): 627-631.

The numbering of squares in a Karnaugh map is not sequential, because the numbering is based on the concept of adjacent states. *Two states are adjacent when they differ in the value of only one variable.* This concept is important because the algebraic combination of two adjacent states results in the elimination of one variable by the axiom $x+x'=1$.

**Figure 201 K Map for one variable**

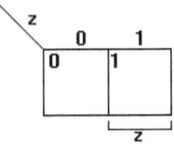

This becomes clear when minterm (square) numbers are in binary format (Figure 205). The 0 1 3 2 sequence is used because it results in adjacent states. Starting from any square, a horizontal or vertical move to any adjacent square changes only one bit in the minterm number. For example $00 \leftrightarrow 01 \leftrightarrow 11 \leftrightarrow 10 \leftrightarrow 00$

A move off the map from squares on the map edges is implemented by wrapping around behind the map to the opposite edge (think of the map as a cylinder). That is to say, a move is from end to end of each row or column. Small brackets on the external edges of a map mark rows and columns where variables w, x, y, z are true.

**Figure 202 K Map with two variables**

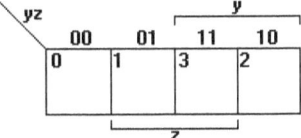

**Figure 203 K Map for three variables**

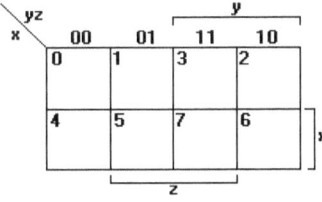

**Figure 204 K Map with four variables**

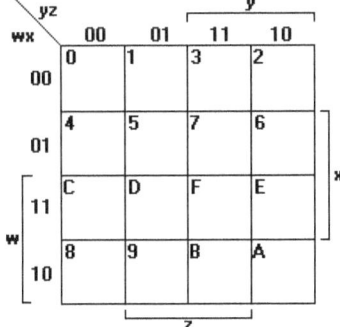

**Figure 205 K Map with binary minterm numbers**

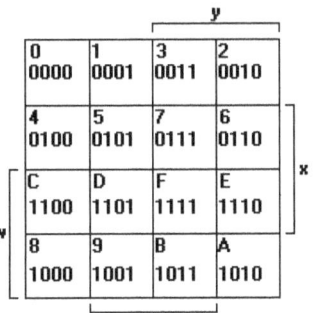

## 2.2 Clusters

Utility of the K map depends upon our ability to recognize certain patterns of 1s or 0s that are known as clusters or subcubes. *Clusters of ones[3] are groups of squares that are adjacent states.* The adjacent property requires that clusters are rectangles or squares, and that the number of squares in a cluster equals a power of 2.

***Two variable map*** A two variable map has four squares (Figure 206). Clusters have 2 or 4 squares. A four square cluster means $F(y,z)=1$.

Two square clusters can occur in four ways. Each of these clusters provide the possibility of eliminating one of four literals[4] y, y', z, z'. A bit drops out if it *changes* when moving to adjacent squares. The functions $f_A$, $f_B$, $f_C$, and $f_D$ correspond to the maps in Figures 206a, b, c, d. Note how the literal *drops out (d)* in each case, because $a+a'=1$. (The + is OR, not addition.)

| | | | | |
|---|---|---|---|---|
| 00 | $f_A = m_0 + m_1$ | 01 | $f_B = m_1 + m_3$ | 11 $f_C = m_3 + m_2$ 10 $f_D = m_2 + m_0$ |

$$00 \quad f_A = m_0 + m_1 \qquad 01 \quad f_B = m_1 + m_3 \qquad 11 \quad f_C = m_3 + m_2 \qquad 10 \quad f_D = m_2 + m_0$$
$$\underline{01} \quad = y'z' + y'z \quad \underline{11} \quad = y'z + yz \quad \underline{10} \quad = yz + yz' \quad \underline{00} \quad = yz' + y'z'$$
$$\underline{0d} \quad = y' \qquad d1 \qquad = z \qquad 1d \qquad = y \qquad d0 \qquad = z'$$

**Figure 206 Two variable K Maps with two square clusters**

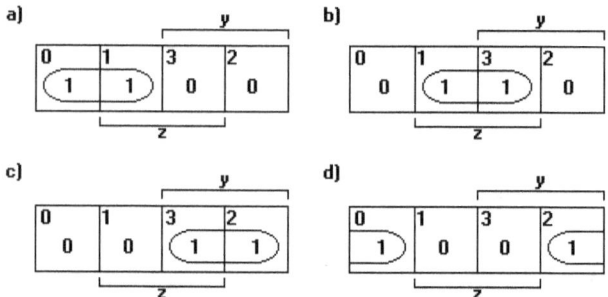

***Three variable map*** A three variable map has eight squares (Figure 207). The map can have clusters of 2, 4, or 8 squares. An eight square cluster means $F(x,y,z)=1$. Two square clusters can occur in each *row* in the four ways shown in Figure 206. In addition two square clusters, such as $m_3/m_7$, can occur four ways in the four *columns* of Figure 207. Each of these two square clusters eliminate one of six literals x, x', y, y', z, z'.

---

[3] Or clusters of zeros
[4] Literal: A variable or the complement of a variable.

For example, literal z drops out of the 0&1 cluster, because 000 & 001 produce x'y'. Here are combinations for four two square clusters in the first row, and two column two square clusters that are identified by their minterms. The three digit numbers, e.g. 010, represent the literals xyz respectively.

| $m_0$ 000 | $m_1$ 001 | $m_3$ 011 | $m_2$ 010 | $m_0$ 000 | $m_3$ 011 |
|---|---|---|---|---|---|
| $m_1$ 001 | $m_3$ 011 | $m_2$ 010 | $m_0$ 000 | $m_4$ 100 | $m_7$ 111 |
| $00d \rightarrow x'y'$ | $0d1 \rightarrow x'z$ | $01d \rightarrow x'y$ | $0d0 \rightarrow x'z'$ | $d00 \rightarrow y'z'$ | $d11 \rightarrow yz$ |

Four square clusters covering a *full* row can occur in two ways (Figure 207a, b). Four square clusters covering two *half* rows can occur in four ways (Figure 207c, d, e, f). Each of these clusters eliminate two literals from the set of six literals x, x', y, y', z, z'. Decimal square numbers are converted to binary to analyze bit changes in adjacent states (not shown here). Again, if a bit *changes* it drops out (*d*) from the expression, because $b+b'=1$.

**Figure 207 Three variable K Maps with four square clusters**

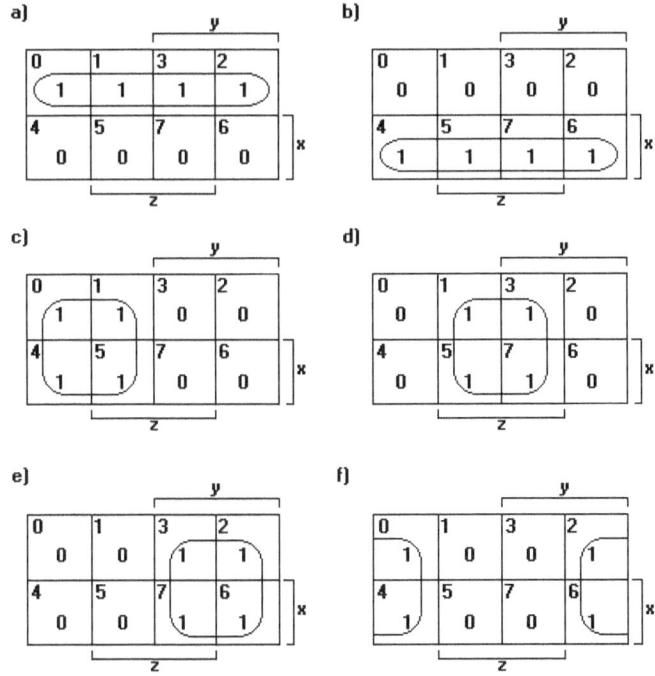

Similar comments apply to four square clusters occupying two rows.

| 207c | 207e | 207f | 207d | 207a | 207b |
|---|---|---|---|---|---|
| $m_0$ 000 | $m_3$ 011 | $m_2$ 010 | $m_1$ 001 | $m_0$ 000 | $m_4$ 100 |
| $m_1$ 001 | $m_2$ 010 | $m_6$ 110 | $m_3$ 011 | $m_1$ 001 | $m_5$ 101 |
| $m_4$ 100 | $m_7$ 111 | $m_0$ 000 | $m_5$ 101 | $m_3$ 011 | $m_7$ 111 |
| $m_5$ 101 | $m_6$ 110 | $m_4$ 100 | $m_7$ 111 | $m_2$ 010 | $m_6$ 110 |
| $d0d \rightarrow y'$ | $d1d \rightarrow y$ | $dd0 \rightarrow z'$ | $dd1 \rightarrow z$ | $0dd \rightarrow x'$ | $1dd \rightarrow x$ |

***Four variable map*** A four variable map has sixteen squares (Figure 208). It can have clusters of 2, 4, 8, or 16 squares. A sixteen square cluster means $F(w,x,y,z)=1$.

Two square clusters can occur in each row in the four ways shown in Figure 206. In addition two square clusters can occur in each of four columns in the same way they can occur in rows (not illustrated). Each of these clusters eliminate one of eight literals w, w', x, x', y, y', z, z'.

Here is an analysis of a sample of 4 clusters (Figure 208). The four digit numbers, e.g. 0100, represent the literals wxyz.

| $m_4$ 0100 | $m_7$ 0111 | $m_9$ 1001 | $m_C$ 1100 |
|---|---|---|---|
| $m_5$ 0101 | $m_F$ 1111 | $m_B$ 1011 | $m_8$ 1000 |
| $010d \rightarrow w'xy'$ | $d111 \rightarrow xyz$ | $10d1 \rightarrow wx'z$ | $1d00 \rightarrow wy'z'$ |

Four square clusters covering whole rows or columns can occur in eight ways. Two ways are shown in Figures 208a, b. Four square clusters covering two half rows can occur in eight ways. Four ways are shown Figures 208c, d, e, f. Each of these clusters eliminate two literals from the set of eight literals w, w', x, x', y, y', z, z'. Here is an analysis of a sample of four square clusters.

| 208a | | | 208e |
|---|---|---|---|
| $m_C$ 1100 | $m_0$ 0000 | $m_1$ 0001 | $m_0$ 0000 |
| $m_D$ 1101 | $m_1$ 0001 | $m_3$ 0011 | $m_2$ 0010 |
| $m_F$ 1111 | $m_4$ 0100 | $m_9$ 1001 | $m_8$ 1000 |
| $m_E$ 1110 | $m_5$ 0101 | $m_B$ 1011 | $m_A$ 1010 |
| $11dd \rightarrow wx$ | $0d0d \rightarrow w'y'$ | $d0d1 \rightarrow x'z$ | $d0d0 \rightarrow x'z'$ |

**Figure 208 Four variable K Maps with four square clusters**

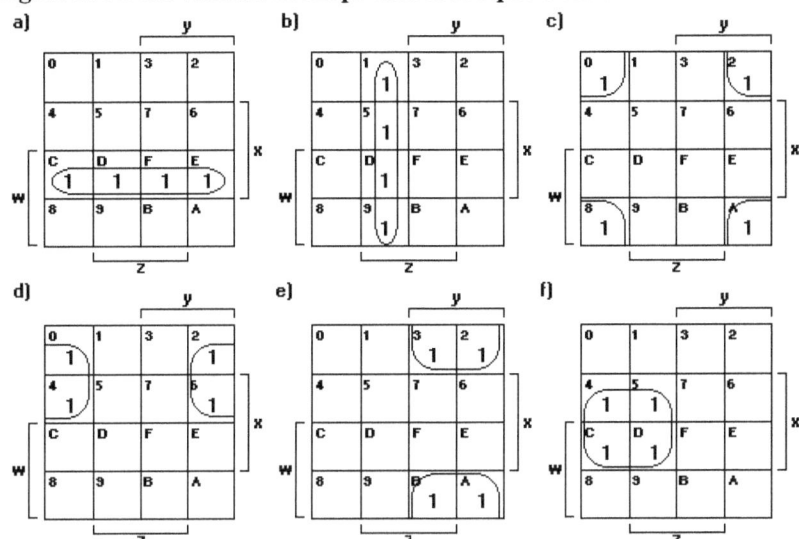

Eight square clusters covering two whole rows or columns can occur in eight ways (six ways are shown Figure 209). Each of these clusters eliminate three literals from the set of eight literals w, w', x, x', y, y', z, z'. Emphasis - variables in columns or rows that change drop out.

| 209($e$) | 209($f$) | 209($c$) | 209($a$) |
|---|---|---|---|
| $m_0$ 0000 | $m_0$ 0000 | $m_0$ 0000 | $m_4$ 0100 |
| $m_1$ 0001 | $m_1$ 0001 | $m_2$ 0010 | $m_5$ 0101 |
| $m_3$ 0011 | $m_4$ 0100 | $m_4$ 0100 | $m_7$ 0111 |
| $m_2$ 0010 | $m_5$ 0101 | $m_6$ 0110 | $m_6$ 0110 |
| $m_8$ 1000 | $m_C$ 1100 | $m_C$ 1100 | $m_C$ 1100 |
| $m_9$ 1001 | $m_D$ 1101 | $m_E$ 1110 | $m_D$ 1101 |
| $m_B$ 1011 | $m_8$ 1000 | $m_8$ 1000 | $m_F$ 1111 |
| $m_A$ 1010 | $m_9$ 1001 | $m_A$ 1010 | $m_E$ 1110 |
| $d0dd \rightarrow x'$ | $dd0d \rightarrow y'$ | $ddd0 \rightarrow z'$ | $d1dd \rightarrow x$ |

33

**Figure 209 Four variable K Maps with eight square clusters**

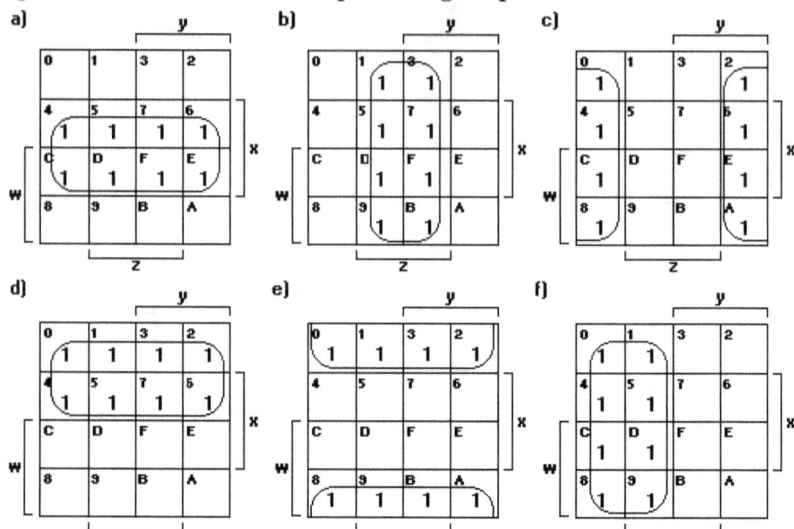

## 2.3 Writing Functions to a Map

A major use of the K map is function simplification. A function is written to the map term by term. Each term belongs to one or more clusters. Then you forget these original clusters, and proceed to select new, hopefully larger clusters to read new terms from the map that add up to a simplified function. This procedure is probably best explained by examples.

***Canonical forms*** If a function is in canonical sum of minterm products (OR of ANDs) format (3.3.1 page 51), then each term in this sum of minterms sets the corresponding K map squares to 1. The process is straightforward, because the square numbers equal the minterm numbers.

If a function is in canonical product of maxterm sums (AND of ORs) format (3.3.2 page 54) then each term in this product of maxterms also sets the corresponding squares to 0. However, the process is not as straightforward, because the maxterm numbers seem harder to perceive. Remember that maxterm primed variables are 1's and not-primed variables are 0's: the maxterm $(x+y'+z')$ has number $011_2$ or $3_{10}$.

However, most functions are not in canonical form.

***Convert the equation***   $F(x, y, z) = xy' + y'z' + x'yz$  *to a K map*

Variable z is not in the term xy' that implies two minterms $m_4$ and $m_5$ (xy'z'+ xy'z) are involved (Figure 210). The xy' term is at the intersection of row x and the two y' columns. Variable x is not in the term y'z'. The y'z' term is at the intersection of x and x' rows and the y'z' column. The intersection contains two minterms: $m_4$ and $m_0$. The term x'yz has all variables in it. This is the xyz minterm $m_3$. This square is at the intersection of column yz and row x'. The function is the sum of minterms $m_0$, $m_3$, $m_4$, and $m_5$.

**Figure 210 K Map for xy'+y'z'+x'yz**

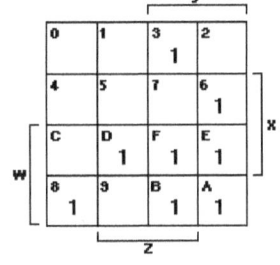

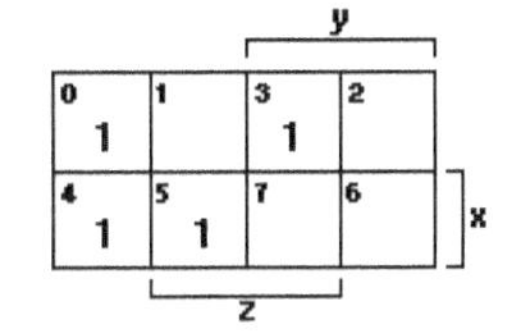

### Practice
Write $f = wy+wxz+w'xyz'+w'x'yz+wx'y'z'$ to a K map (Figure 211)

Write $f = wxz+w'xz+wxz'+wx'y'z$ to a K map (Figure 212)

Write $f = w'x'z'+ w'xy +wxz+w x'y'$ to a K map (Figure 213)

**Figure 211**

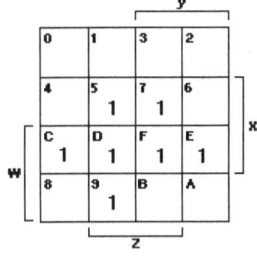


**Figure 212**

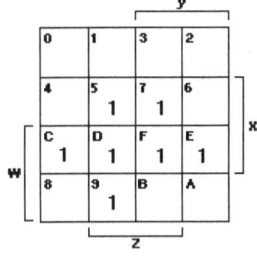

**Figure 213**

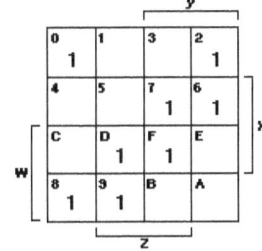

---

*Problems* 201-204 Plot each pair f and g on the same K map.

201 $f = (yz)'$          $g = yz$
202 $f = y + z$          $g = y'z'$
203 $f = y \ xor \ z$     $g = y \ xnor \ z$
204 $f = xy + x'z + yz$   $g = (x' + y')(x + z')(y' + z')$

## 2.4 Reading Functions from a Map

Reading a map is a matter of selecting clusters with a square or rectangular shape that includes a number of squares equal to a power of 2. The goal is to simplify the function by finding smallest number of clusters that include all the 1's or all of the 0's.

***Reminder*** Since $x + x = x$ (Table 3.1 Idempotent 5 page 45) any square can be in more than one cluster.

***Selecting clusters*** Start the process by noting those 1 entries in the map that can be included in only one larger cluster. Include these unique 1's in the largest possible clusters. Select the clusters that include the greatest number of squares with 1's. Take it from there. Keep in mind that some clusters may be hard to perceive, and that there may be alternative ways to form clusters.

***Reading the map four ways*** (Figure 210)

1 outputs sum of products    $f = m_0 + m_3 + m_4 + m_5$

0 outputs sum of products    $f' = m_1 + m_2 + m_6 + m_7$

1 outputs products of sums    $f' = M_0 M_3 M_4 M_5$

0 outputs products of sums    $f = M_1 M_2 M_6 M_7$

***Read the map 1's in Figure 211***

The 4-cluster wy stands out (squares F,E,B,A)

The 2-cluster in column yz' is has third variable x, so the term is xyz' (squares 6,E).

The 2-cluster in column yz is has third variable x', so the term is x'yz (squares 3,B).

The 2-cluster in row wx is has third variable z, so the term is wxz (squares D,F)

The 2-cluster in row wx' is has third variable z', so the term is wx'z' (squares 8,A). Observe that the 2-clusters cover the 4-cluster wy so that it is redundant. Therefore sum of the terms is $f = xyz' + x'yz + wxz + wx'z'$.

**Figure 210 K Map**
**for xy'+y'z'+x'yz**

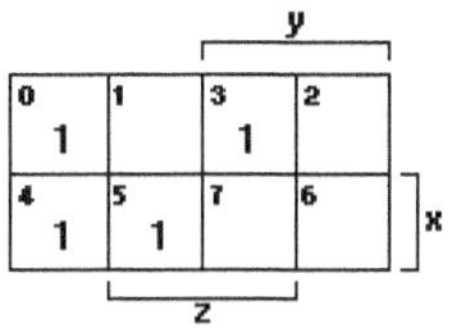

**Figure 211 K Map**

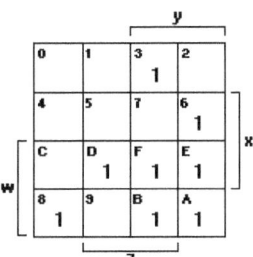

### Read the map 1's in Figure 212

The 4-cluster in row wx drops out variables y and z, so the term is wx (C,D,F,E). The 4-cluster at the intersection of rows x and columns z is xz (5,7,D,F). This overlaps two squares in row wx, and variables w, y drop out. The 2-cluster in column y'z is has third variable w, so the term is wy'z (D,9). This overlaps one square in row wx. The sum of the terms is $f = wx + xz + wy'z$.

**Figure 212 K Map**

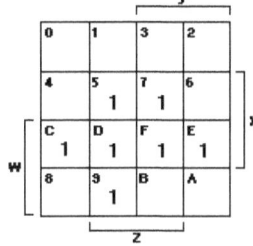

**Figure 213 K Map**

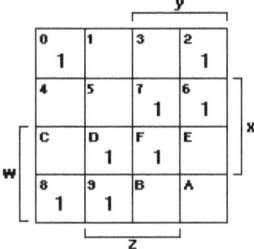

### Read the map 1's in Figure 213 This time we read column clusters.

The 2-cluster in column yz' is has third variable w', so the term is w'yz' (2,6). The 2-cluster in column yz is has third variable x, so the term is xyz (7,F). The 2-cluster in column y'z is has third variable w, so the term is wy'z (D,9). The 2-cluster in column y'z' is has third variable x', so the term is x' y'z' (0,8).

The sum of the terms is $f=w'yz'+xyz+wy'z+x'y'z'$. This equation by columns differs in form from the equation read when clustering by rows: $f=w'x'z'+w'xy+wxz+wx'y'$. This illustrates some of the effects of different cluster selections.

Digital Design

## Read the map 0's in Figure 212

If you pretend the zeros (Figure 212) are ones then you can get the complement of the function $f$ as a sum of terms. When $f$ is complemented the result is $f$ in the form of a product of terms. This process is less prone to error than trying to write down the maxterms directly. Try it.

The 4-cluster in row w'x' is covered by the term w'x' (0,1,2,3).
The 4-cluster at the four corners is x'z' (0,2,8,A).
The 2-cluster at the intersection of columns y and row wx' is wx'y (A,B).
The 4-cluster in rows w' has second variable z', so the term is w'z' (0,2,4,6).

The sum of the terms is $f' = w'x' + x'z' + wx'y + w'z'$. Now find $f$ by taking the complement.

$$f = (f')' = (w'x' + x'z' + wx'y + w'z')'$$
$$f = (w+x)(x+z)(w'+x+y')(w+z)$$

*multiply out to get the original sum of terms*

$$f = (w+x)(x+z)(w'+x+y')(w+z)$$
$$= (w+x)(w+z)(x+z)(x+w'+y')$$
$$= (w+xz)(x+zw'+zy') \qquad\qquad \text{See Figure 212}$$
$$= wx+wzw'+wzy'+xzx+xzw'+xzy'$$
$$= wx+0+wzy'+xz(1+w'+y')$$
$$= wx+xz+wy'z$$

---

*Problems* 205-215 Plot the functions on K maps. Use map entered variables for 205-210 (page 40).

205 $d_2 = S_3 \times kbd + S_4 \times kbd'$ (from page 10)

206 $d_1 = S_1 + S_3 \times kbd'$

207 $d_0 = S_0 \times (R < N) \times (R = 0)' + S_3 \cdot kbd'$

208 $d_0 = q_0$ xor $p$

209 $d_1 = q_1$ xor $pq_0$

210 $d_2 = q_2$ xor $pq_1q_0$

211 $f = z(z' + y)$

212 $f = xy + x'z + yz$

213 $f = (x + y)(x' + z)(y + z)$

214 $f = x'yz' + w'(y + z)$

215 $f = xz' + y(wx + w'x' + y')$

---

## 2.5 Don't Care Conditions

A minterm that is not specified is referred to as a don't care term. A dash is placed in the corresponding K map square (Figure 214). When you read the map you assign 0 or 1 to each don't care term. Your guide is the desire to maximize cluster sizes. However the logic of the project may prevent you from maximizing cluster sizes.

### *Read the map 1's in Figure 214*

Suppose the dash in row wx is assigned a 1 (Figure 214), then the 4-cluster in row wx is independent of y and z, so the term is wx (C, D, F, E).

If the dash in square 4 is assigned a 1, then the 4-cluster at the intersection of rows x and columns z' produces the term xz' (4, 6, C, E).

The 2-cluster at the intersection of column yz' and rows w is wyz' (E, A). The remaining dashes are assigned zeros, because they have not been included in any clusters. The sum of the terms is $f = wx + xz' + wyz'$

**Figure 214 K Map with don't care terms**

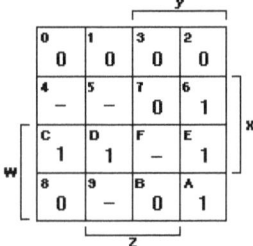

## 2.6 AND and OR operations on Maps

If you need the logical OR of two functions you can OR their K maps. Simply OR the values of the minterms in each square. If you need the logical AND of two functions you can AND their K maps. Simply AND the values of the minterms in each square.

**Figure 215 OR of two K Maps**

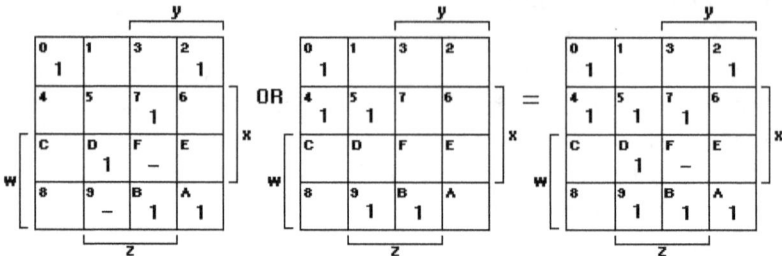

**Figure 216 AND of two K Maps**

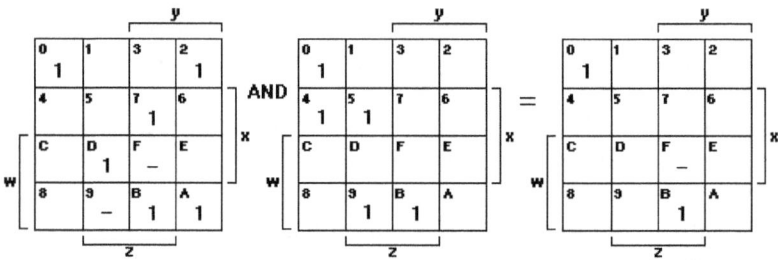

## 2.7 Map entered Variables

The number of squares in a K map is doubled each time an input variable is added. The number of squares is halved when an input variable is converted to a map entered variable. The K map is massive when there are more than four variables, and then the K map is difficult to work on. The number of squares is 2, 4, 8, and 16 for 1, 2, 3, and 4 input variables.

***Basic idea*** The coefficient of a term such as xyz in an algebraic expression is an implied 1. In any algebra any part of a term multiplied by the remaining part is defined as a coefficient of the remaining part. For the term xyz, the coefficient of yz is x, the coefficient of y is xz, and so forth. When K map entries are limited to 1, 0, and −, then the variables *jkq* are the K

**Figure 217 K map without a map entered variable**

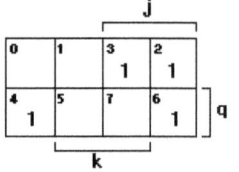

map coordinates in this example (Figure 217). When an input variable is converted to a map entered variable then that variable is removed from the K map coordinates to halve the map's size.

### Write $f = jq' + k'q$ to a map with variables j, k, and q.

The term jq' maps to the intersection of row q' and 2 columns j (3, 2 Figure 217). The term k'q maps to the intersection of row q and two columns k' (4, 6).

### Write $f = q'j + qk'$ to a map with variables j and k

The equation $f = jq' + k'q$ has three variables: j, k, and q. The number of K map variables reduces to two if q is treated as a coefficient.
$$f = jq' + k'q = q'(j) + q(k')$$

• Term **j** represents the cluster consisting of minterms $m_2$ and $m_3$. Enter **q'** in squares 2 and 3 (Figure 218a).

• Term **k'** represents the cluster consisting of minterms $m_0$ and $m_2$. Enter **q** in squares 0 and 2. The coefficient $q' + q$ in square 2 equals 1 (Fig 218b).

### Reading a map with map entered variables

A minterm coefficient is the value in the square representing the minterm The function read from the map in Figure 218 is produced as follows.

$$f = q(m_0) + 0(m_1) + 1(m_2) + q'(m_3)$$
$$= q(j'k') + 0(j'k) + 1(jk') + q'(jk)$$
$$= qj'k' + 0 + (q+q')jk' + q'jk$$
$$= q(j'k' + jk') + q'(jk' + jk)$$
$$= qk' + jq'$$

**Figure 218 K map with a map entered variable**

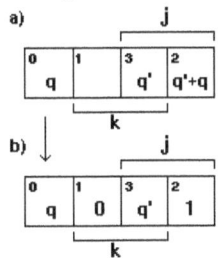

Read the K map in Figure 219 so that the coefficients of the minterms are the values in the squares representing the minterms.

$$f = wr'm_0 + t'm_2 + tm_4 + m_5 + m_6$$
$$f = wr'x'y'z' + t'x'yz' + txy'z' + xy'z + xyz'$$

**Figure 219**

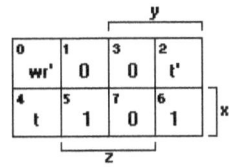

---

*Problems* 216-218 Make K maps with map entered variables (mev).

216 *mux* $f(y,z) = yza_3 + yz'a_2 + y'za_1 + y'z'a_0$    (mev $a_0$ to $a_3$)

217 *Full adder* $s = x \oplus y \oplus c_{in}$   $c = xy + c_{in}(x+y)$   (mev $c_{in}$)

218 $d_2 = q_2 \oplus pq_1q_0$   (mev p)

---

# Review 2

***The Karnaugh Map*** A Karnaugh map (K map) is a graphical display of a switching function. The map is an array of squares where each square is a state. The Karnaugh Map is based on the concept of *adjacent* states. Two states are adjacent when they differ in the value of only one variable. The function's input variables are the array's coordinates. The function's output is shown as 0 or 1 in the squares. Each square of a K map is equivalent to a truth table row. This is why each square represents a minterm or a maxterm. The squares are numbered. Square n corresponds to truth table row n. Each square is assigned the switching function 0 or 1 output value of the corresponding row in a truth table.

***Clusters*** Utility of the K map depends upon the ability to recognize certain patterns of 1s or 0s that are known as clusters. Clusters represent terms of the map's function. Clusters are groups of squares that are adjacent states. The adjacent property requires that clusters are rectangles or squares, and that the number of squares in a cluster is a power of 2.

***Writing Functions to a Map, Reading Functions from a Map*** A major use of the K map is function simplification. A function is written to the map term by term. Each term adds a cluster of 1, 2, etc. squares. Then we forget the original clusters, and proceed to select new, hopefully larger, clusters to read new terms from the map that add up to a simplified function.

***Don't Care Conditions*** A minterm that is not specified is referred to as a don't care term. A dash is placed in the corresponding K map square. When one reads the map one assigns 0 or 1 to each don't care term.

***AND and OR operations on Maps*** If one needs the logical OR of two functions one can OR their K maps. Simply OR the values of the minterms in each pair of squares. If one needs the logical AND of two functions one can AND their K maps. Simply AND the values of the minterms in each pair of squares.

***Map entered Variables*** The number of squares is 2, 4, 8, and 16 for 1, 2, 3, and 4 input variables. The number of squares in a K map is doubled each time an input variable is added. The number of squares is halved when an input variable is converted to a map entered variable.

# 3 Switching Algebra

Switching algebra is the theoretical basis for digital logic design.

Switching algebra provides a rigorous basis for analysis and synthesis of logic functions. Truth Tables and Karnaugh Maps add a visual dimension to analysis of logic functions. Their use simplifies analysis and design of logic functions.

Digital circuits process signals which are in one of two states at any instant of time. Any signal voltage is high (H) or low (L). In a physical system the voltage levels H and L (High and Low) at nodes correspond to the Boolean logic values 1 and 0 (true and false) or vice versa, or both (at different nodes). Binary digits 0 and 1 are referred to as bits. (The word switching is a carryover from the days when systems were assembled with relays.) C. E. Shannon[1] was one of the first people to apply mathematical methods to switching theory.

A switching algebra is a special case of Boolean algebra[2]. As in Euclid's geometry and other algebra's, the switching algebra is based on a set of axioms. The set of axioms used here were created by E. V. Huntington[3]. The axioms are used to prove theorems. The axioms and newly proven theorems are used to prove more theorems, and so on. This is followed by a systematic study of the operations and the rules of switching algebra arithmetic. Samuel H. Caldwell wrote what seems to be the first logical design text[4] that is very well written.

The equations produced by use of switching algebra have two canonical forms that are a sum of products (minterms) or a product of sums (maxterms). The canonical forms are important, because they facilitate use of two design tools, the truth table and the Karnaugh map. We show how functions are represented as canonical forms.

---

[1] Shannon, C. E. A Symbolic Analysis of Relay and Switching Circuits. *Transactions AIEE* 57 (1938): 713-723.
[2] Boole, G. *An Investigation of the Laws of Thought.* ISBN 0486 600 289
[3] Huntington, E. V. Sets of Independent Postulates for the Algebra of Logic. *Transactions Am. Math. Soc.* 5 (July 1904): 288-309.
[4] Caldwell, S. H. Switching Circuits and Logic Design. (1958) ISBN 1124 074 937

Digital Design

# 3.1 Axioms for a Switching Algebra

Huntington. proved that the following six axioms form the basis for a Boolean algebra. The operators $+$, $\bullet$, ' are known as OR, AND, and NOT.

**Axiom 1**. For any member x of set S, members 0 and 1 of S have the properties

$$x + 0 = x \qquad (1)$$
$$x \bullet 1 = x \qquad (1d)$$

**Axiom 2**. All members x, y of S commutate.

$$x + y = y + x \qquad (2)$$
$$x \bullet y = y \bullet x \qquad (2d)$$

**Axiom 3**. All members x, y, z of S distribute.

$$x \bullet (y + z) = (x \bullet y) + (x \bullet z) \qquad (3)$$
$$x + (y \bullet z) = (x + y) \bullet (x + z) \qquad (3d)$$

**Axiom 4**. All members x of S have an inverse member x'.

$$x + x' = 1 \qquad (4)$$
$$x \bullet x' = 0 \qquad (4d)$$

**Axiom 5**. At least two members x,y of S are not equal.

**Axiom 6**. If x and y are members of S then $x \bullet y$ and $x+y$ are members of S.

**Operator Precedence**: execute ', then $\bullet$, then $+$.

**Duals** In the axioms an equation and its dual are shown. A dual is formed when 0 is replaced by 1, 1 is replaced by 0, $+$ is replaced by $\bullet$, and $\bullet$ is replaced by $+$. Consider axiom 4:

The dual of $x + x' = 1$ is formed by substituting $\bullet$ for $+$ and 0 for 1 to get $x \bullet x' = 0$. *If the equation number is 4, then the dual number is 4d.*

This implements the principle of duality. If a Boolean statement is true, then the dual of the statement is true. The principle is applicable to any theorem in switching algebra.

## Table 3.1 Axioms and theorems of Switching Algebra

Operations with 1 and 0: axiom

1. $x + 0 = x$        1d. $x \cdot 1 = x$

Commutative: axiom

2. $x + y = y + x$        2d. $x \cdot y = y \cdot x$

Distributive: axiom

3. $x(y + z) = xy + xz$        3d. $x + yz = (x + y)(x + z)$

Complementarity: axiom

4. $x + x' = 1$        4d. $x \cdot x' = 0$

Idempotent:

5. $x + x = x$        5d. $x \cdot x = x$

Involution:

6. $(x')' = x$

Complementarity:

7. $x + 1 = 1$        7d. $x \cdot 0 = 0$

Consensus:

8.   $xy + x'z + yz = xy + x'z$

8d   $(x + y)(x' + z)(y + z) = (x + y)(x' + z)$

Simplification:

9.   $x + xy = x$        9d.   $x(x + y) = x$

10. $xy + xy' = x$        10d. $(x + y)(x + y') = x$

11. $x + x'y = x + y$        11d. $x(x' + y) = xy$

Lemma 1.

12.    $x + (x' + y) = 1$      12d   $x(x'y) = 0$

DeMorgan:

13.    $(x + y)' = x'y'$   and for n variables $(x + y + ... + n)' = x'y'.....n'$

13d.   $(xy)' = x' + y'$   and for n variables $(xy.....n)' = x' + y' + ... + n'$

Associative:

14     $x + (y + z) = (x + y) + z$      14d   $x(yz) = (xy)z$

## 3.2 Theorems

The two methods for proving theorems are an algebraic process, and the method of perfect induction (3.3.3 page 56). The first theorems we prove use the axioms as a basis for proof. The proof of subsequent theorems are based on axioms, and theorems just proved. The proofs may be considered as examples showing how switching algebra can be applied. We believe this is the best reason that working through them is worth while.

**Theorem: The member 0 is unique.**
To any member x apply Axiom 1 when there are two zero members 0 & p.

$x{+}0 = x$    *and*    $x{+}p = x$

Substitute p for x in the first equation and 0 for x in the second

$p{+}0 = p$   *and*   $0{+}p = 0$

Apply Axiom 2:

$p = p{+}0 = 0{+}p = 0$    $\Rightarrow$    $p = 0$

**Theorem: The member 1 is unique.**
To any member x apply Axiom 1d when there are 2 one members 1 & p.

$x{\cdot}1 = x$    *and*    $x{\cdot}p = x$

Substitute p for x in the first equation and 1 for x in the second

$p{\cdot}1 = p$    *and*    $1{\cdot}p = 1$

Apply Axiom 2d:

$p = p{\cdot}1 = 1{\cdot}p = 1$    $\Rightarrow$    $p = 1$

**Theorem: $x + x = x$ for all x in the set S.**    **(5)**

$$
\begin{aligned}
Proof \quad x + x &= (x + x)(1) & (1d)\\
&= (x + x)(x + x') & (4)\\
&= x + xx' & (3d)\\
&= x + 0 & (4d)\\
&= x & (1)
\end{aligned}
$$

**Theorem: $x \bullet x = x$ for all x in the set S.**    **(5d)**

$$
\begin{aligned}
Proof : \quad x \cdot x &= xx + 0 & (1)\\
&= xx + xx' & (4d)\\
&= x(x + x') & (3)\\
&= x(1) & (4)\\
&= x & (1d)
\end{aligned}
$$

**Theorem:** $(x')' = x$ **for all x in the set S.**    **(6)**

*Proof* :

*If* $x'$ *is the complement of* $x$ *(i.e. the inverse element of* $x$*), then*

$x + x' = 1$  (4) *and* $x \cdot x' = 0$   (4d)

*If* $(x')'$ *is the complement of* $x'$ *then*

$x' + (x')' = 1$  (4) *and* $x' \cdot (x')' = 0$   (4d)

*so that* $x + x' = x' + x = 1 = x' + (x')'$   $\Rightarrow$   $x = (x')'$

**Theorem:** $x + 1 = 1$ **for all x in the set S.**    **(7)**

$$
\begin{aligned}
\textit{Proof}: \quad x + 1 &= (x+1)(1) & (1d) \\
&= (x+1)(x+x') & (4) \\
&= x + 1x' & (3d) \\
&= x + x' = 1 & (1d)(4)
\end{aligned}
$$

**Theorem:** $x \cdot 0 = 0$ **for all x in the set S.**    **(7d)**

$$
\begin{aligned}
\textit{Proof}: \quad x \cdot 0 &= (x \cdot 0) + (0) & (1) \\
&= (x \cdot 0) + (x \cdot x) & (4d) \\
&= x(0 + x') & (3) \\
&= xx' = 0 & (1)(4d)
\end{aligned}
$$

**Theorem:** $xy + x'z + yz = xy + x'z$ **for all x,y in the set S. (8)**

$$
\begin{aligned}
\textit{Proof}: \quad xy + x'z + yz &= xy + x'z + yz \cdot 1 & (1d) \\
&= xy + x'z + yz(x + x') & (4) \\
&= xy(1 + z) + x'z(1 + y) & (3)(1d) \\
&= xy + x'z & (7)
\end{aligned}
$$

**Theorem:** $(x+y)(x'+z)(y+z) = (x+y)(x'+z)$ **for all x,y in the set S.   (8d)**

$$
\begin{aligned}
\textit{Proof}: \quad (x+y)(x'+z)(y+z) &= (x+y)(x'+z)(y+z+0) & (1) \\
&= (x+y)(x'+z)(y+z+xx') & (4d) \\
&= (x+y)(x'+z)(y+z+x)(y+z+x') & (3d) \\
&= (x+y+0)(x'+z+0)(y+z+x)(y+z+x') & (1) \\
&= (x+y+0 \cdot z)(x'+z+0 \cdot y) & (3d) \\
&= (x+y+0)(x'+z+0) & (7d) \\
&= (x+y)(x'+z) & (1)
\end{aligned}
$$

**Theorem:** $x + xy = x$ *for all x,y in the set S.*     **(9)**
Proof :     $x + xy = x \cdot 1 + xy$          $(1d)$
$\qquad\qquad = x(1 + y)$          $(3)$
$\qquad\qquad = x(1)$          $(7)$
$\qquad\qquad = x$          $(1d)$

**Theorem:** $x(x + y) = x$ *for all x,y in the set S.*     **(9d)**
Proof :     $x(x + y) = xx + xy$          $(3)$
$\qquad\qquad = x + xy$          $(5d)$
$\qquad\qquad = x$          $(9)$

**Theorem:** $xy + xy' = x$ *for all x,y in the set S.*     **(10)**
Proof :     $xy + xy' = x(y + y')$          $(3)$
$\qquad\qquad = x(1)$          $(4)$
$\qquad\qquad = x$          $(1d)$

**Theorem:** $(x+y)(x+y') = x$ *for all x,y in the set S.*     **(10d)**
Proof :     $(x + y)(x + y') = x + yy'$          $(3d)$
$\qquad\qquad = x + 0$          $(4d)$
$\qquad\qquad = x$          $(1)$

**Theorem:** $x + x'y = x + y$ *for all x,y in the set S.*     **(11)**
Proof :     $x + x'y = (x + x')(x + y)$          $(3d)$
$\qquad\qquad = (1)(x + y)$          $(4)$
$\qquad\qquad = x + y$          $(1d)$

**Theorem:** $x(x'+y) = xy$ *for all x,y in the set S.*     **(11d)**
Proof :     $x(x' + y) = xx' + xy$          $(3)$
$\qquad\qquad = 0 + xy$          $(4d)$
$\qquad\qquad = xy$          $(1)$

**Lemma 1:**     $x + (x' + y) = 1$          $x(x'y) = 0$

The associative property is derived from the axioms. Until proved, the associative property cannot be used in a proof. This lemma simplifies the proofs of DeMorgan's theorems and the associative property by factoring out common parts. In what follows note how parentheses are *not* dropped.

**Lemma:** $x + (x'+ y) = 1$ **for all x,y in the set S.** **(12)**

$$Proof: \quad x+(x'+ y) =1\cdot[x+(x'+ y)] \qquad (1d)$$
$$= (x+ x')[x+(x'+ y)] \qquad (4)$$
$$= x+ x'(x'+ y) \qquad (3d)$$
$$= x+ x' = 1 \qquad (9d)(4)$$

**Lemma:** $x(x'y) = 0$ **for all x,y in the set S.** **(12d)**

$$Proof: \quad x(x'y) = 0+ x(x'y) \qquad (1)$$
$$= xx' + x(x'y) \qquad (4d)$$
$$= x(x'+ x'y) \qquad (3)$$
$$= xx' = 0 \qquad (9)(4d)$$

**Theorem:** $(x+y)' = x'y'$ **for all x,y in the set S.** **(13)**
**and for n variables** $(x+y+...+n)' = x'y'...n'$ **(DeMorgan)**

$Method: \quad let \ r = (x+ y) \ and \ \ s = (x'y')$
$If \ r + s = 1, and \ rs = 0, then \ r' = s \ by \ axiom \ (4)$
$$Proof: \quad (x+ y)+(x'y') = [(x+ y)+ x']\cdot[(x+ y)+ y'] \quad (3d)$$
$$= 1\cdot 1 \qquad (12)$$
$$= 1 \qquad (1d)$$
$$(x+ y)\cdot(x'y') = x(x'y')+ y(yx') \qquad (3)(2d)$$
$$= 0\cdot 0 = 0 \qquad (12d)(1d)$$
$Therefore \quad (x+ y)' = x'y'$

**Theorem:** $(xy)' = x'+y'$ **for all x,y in the set S.** **(13d)**
**and for n variables** $(xy...n)' = x'+y'+...+n'$ **(DeMorgan)**

$Method: \quad let \ r = (x+ y) \ and \ \ s = (x'y')$
$If \ r + s = 1, and \ rs = 0, then \ r' = s \ by \ axiom \ (4)$
$$Proof: \quad (xy)+(x'+ y') = [x+(x'+ y')]\cdot[y+(x'+ y')] \quad (3d)$$
$$= 1\cdot 1 \qquad (12)$$
$$= 1 \qquad (1d)$$
$$(xy)\cdot(x'+ y') = (yx)x'+(xy)y' \qquad (3)(2d)$$
$$= 0\cdot 0 = 0 \qquad (12d)(1d)$$
$Therefore \quad (xy)' = x'+ y'$

**Theorem:** *$x+(y+z) = (x+y)+z$ for all x,y in the set S.* **(14)**

*Method : let $r = (x+y)+z$, $s = x+(y+z)$*
*If $r+s'' = 1$, and $rs' = 0$  $s' = x'(y'z')$, then $r' = s'$ by axiom (4) & so $r = s$*
*Proof that $r+s' = 1$*       $s' = x'(y'z')$                  (13)

$$r+s' = r+x'(y'z')$$
$$= [r+x'][r+(y'z')] \quad (3d)$$
$$= [r+x'][r+y'][r+z'] \quad (3d)$$

*Evaluate each term:*

$r+x' = (x+y)+z+x'$                     $r+y' = 1$                      *same as $r+x'$*
$= (x+y)+x'+z$  (2d)          $r+z' = (x+y)+z+z'$
$= 1+z$        (12)               $= (x+y)+1$     (4)
$= 1$         (7)                $= 1$        (7)

*And so  $r+s' = [r+x'][r+y'][r+z'] = [1][1][1] = 1$*

*Proof that $rs' = 1$*   $rs' = [(x+y)+z]s'$
$= (x+y)s'+zs'$  (3)
$= xs'+ys'+zs'$   (3)

*Evaluate each term:*

$xs' = xx'(y'z') = 0$          (13)(12d)
$ys' = yx'(y'z')$            (13)
$= yy'+yx'(y'z')$         (1)(4d)
$= y[y'+x'(y'z')]$        (3)
$= y[y'+x'][y'+(y'z')]$  (3d)
$= y[y'+x'][y']$         (9)
$= y[y'][y'+x']$         (2d)
$= 0$               (4d)
$zs' = zx'(y'z')$
$= x'z(y'z') = 0$         (2d)(12d)

*Therefore  $rs' = zs'+ys'+zs' = 0+0+0 = 0$*
*And so $x+(y+z) = (x+y)+z$*

(The proof of 14d that $x(yz)=(xy)z$ is similar to the proof for 14c.)

## 3.3 Minterms and Maxterms

A switching function of n variables $F(x,y,...)$ is an expression representing logical decisions. The logic designer perceives the variables $x,y,...$ as inputs to a system, and the function $F(x,y,..)$ as a system output. A system may have more than one output. Each output is represented by a function such as $F$.

Functions are algebraic - that is to say they are formed from variables, and the arithmetic operators tick ('), plus (+), and mul (• or ×) representing the logical operators NOT ('), OR (+), and AND (• or ×). The functions are the end product of the theoretical design of a system. One practical question that arises is *can these functions be simplified?* The answer is sometimes. The inference is that simplified functions have fewer terms. In turn this implies fewer components in the system. A more subtle question is *can the expression of the function be converted into a form that can be physically realized more effectively?* The answer is usually.

*Literal* A variable x may appear in expressions as x or its complement x'. The term literal is introduced to simplify discussions. If you see the term xy'z in an expression it is easier to say *the term is a product of literals x, y, z* than to say *the term is a product of variable x, complemented variable y, and variable z.*

   *Literal represents a variable or the complement of a variable.*

### 3.3.1 Sum of Minterms

Suppose a function $F(x,y,z)$ is a sum of terms, and each term is a product of one or more literals. Expand $F$ so that all terms include every variable.
$$F(x, y, z) = xyz + z + yz = xyz + (x + x')(y + y')z + (x + x')yz$$
$$= xyz + xy'z + x'yz + x'y'z = m_7 + m_5 + m_3 + m_1 = z$$

Now expand this $F(x,y)$ into a sum of terms that include every variable.
$$F(x, y) = x + y$$
$$= x \cdot 1 + y \cdot 1$$
$$= x(y + y') + y(x + x')$$
$$= xy + xy' + yx + yx' = xy + xy + xy' + yx'$$
$$= xy + xy' + x'y = m_3 + m_2 + m_1$$

Digital Design

The algebraic manipulations converted F(x,y,z) and F(x,y) into *sums of minterms.*

A *minterm* is a special product of literals (variables). Special in the sense that *all* variables such as x, y, z, that are in a function, appear in *every* term of any minterm. To repeat - every variable appears in every minterm as a literal.

A term that is a product of literals is a *canonical product term* when there is exactly one occurrence of *every* function variable in the term. Every variable or its complement is used in the term. Every is the key word. A canonical product term has the special name *minterm.* Lower case m is used for the minterm symbol.

> The prefix *min* is used because each minterm is an AND.
> An AND equals 1 for *only one* combination of variables.

For n variables there are $2^n$ minterms. For example, two variables have four minterms. The symbols for the four minterms are $m_0$, $m_1$, $m_2$, and $m_3$.

We can calculate minterm values from a truth table listing all input x,y combinations. Given x and y values the x' and y' values are known. The output minterm columns are filled by multiplying the values of two literals shown at each column head.

| *Variables* | | | | | *Minterms* | | | |
|---|---|---|---|---|---|---|---|---|
| Row | x | y | x' | y' | x'y' | x'y | xy' | xy |
| | | | | | $m_0$ | $m_1$ | $m_2$ | $m_3$ |
| 0 | 0 | 0 | 1 | 1 | 1 | 0 | 0 | 0 |
| 1 | 0 | 1 | 1 | 0 | 0 | 1 | 0 | 0 |
| 2 | 1 | 0 | 0 | 1 | 0 | 0 | 1 | 0 |
| 3 | 1 | 1 | 0 | 0 | 0 | 0 | 0 | 1 |

*Easier way:* You avoid the algebraic manipulations if, for example, you write any function *F* to a K-map. Each 1 in a map square represents a minterm. The function is the sum of the minterms. Write the above F(x,y,z) and F(x,y) to K maps. Then read out the sum of minterms from each map.

Since a 1 in a map square represents a minterm then so does an output 1 in a truth table row. What is the a sum of minterms that equal this $F(x,y)$?

| Row | x | y | F(x, y) | |
|---|---|---|---|---|
| 0 | 0 | 0 | 0 | |
| 1 | 0 | 1 | 1 | $F(x, y) = m_1 + m_2 + m_3 = x'y + xy' + xy = x + y$ |
| 2 | 1 | 0 | 1 | |
| 3 | 1 | 1 | 1 | |

An equation which is the sum of one or more minterms, is in a *sum of products canonical form*. Any term can be converted into canonical form by the use of Axiom 4 ($x + x' = 1$). Other axioms may be involved, but Axiom 4 is the essential one as shown in the following examples.

*Example 3.1 Function is a sum of minterms that can be simplified*

| Row | x | y | F(x, y) |
|---|---|---|---|
| 0 | 0 | 0 | 1 |
| 1 | 0 | 1 | 1 |
| 2 | 1 | 0 | 0 |
| 3 | 1 | 1 | 0 |

$$F(x, y) = m_0 + m_1 = x'y' + x'y = x'(y' + y) = x' \times 1 = x'$$

*Example 3.2 Expand x' into canonical sum of minterms form*
$$F_2(x, y) = x'$$
$$= x' \times 1$$
$$= x'(y + y')$$
$$= x'y + x'y'$$
$$= m_1 + m_0$$

*Example 3.3 Expand 1 into canonical sum of minterms form*
$$F_2(x, y) = 1$$
$$= 1 \times 1$$
$$= (x + x')(y + y')$$
$$= xy + xy' + x'y + x'y'$$
$$= m_3 + m_2 + m_1 + m_0$$

## 3.3.2 Product of Maxterms

Suppose a function F is a product of terms, and each term is sum of literals. For example: $F(x, y, z) = (x + y + z)(z)(y + z)$. A term that is a sum of literals consists of one literal, or the sum of two or more literals. Expand F so that every term includes every variable.

$$F(x, y) = (x + y)(x)(y) = (x + y)(x + 0)(0 + y) = (x + y)(x + yy')(xx' + y)$$
$$= (x + y)(x + y)(x + y')(x + y)(x' + y)$$
$$= (x + y)(x + y')(x' + y) = M_0 M_1 M_2$$
$$check \ F = (x + y)(xx' + xy + y'x' + y'y) = (x + y)(xy + y'x')$$
$$= xxy + xy'x' + yxy + yy'x' = xy + 0 + xy + 0 = xy = m_3$$
$$original \ F = (x + y)(x)(y) = xxy + yxy = xy + yx = xy \quad qed$$

*Reminder for maxterms: ticks are 1's ($y+z' \to 01 \to M_1$). (see page 23)*

The algebraic manipulations converted *F* to a product of maxterms.

A *maxterm* is a special sum of literals (variables). Special in the sense that *all* variables such as x, y, z, that are in a function, appear in *every* term of any maxterm. To repeat - every variable appears in every maxterm as a literal.

A term that is a sum of literals is a *canonical sum term* when there is exactly one occurrence of *every* function variable in the term. Every variable or its complement is used in the term. Every is the key word. A canonical sum term has the special name *maxterm*. Upper case M is used for the maxterm symbol.

> The prefix *max* is used because each minterm is an OR.
> An OR equals 1 for *all but one* combination of variables.

For n variables there are $2^n$ maxterms. For example, two variables have four maxterms. The symbols for the four maxterms are $M_0$, $M_1$, $M_2$, $M_3$.

We can calculate maxterm values from a truth table listing all input x,y combinations. Given x and y values the x' and y' values are known. The output maxterm columns are filled by the values of two literals shown at each column head.

| Variables | | | | | Maxterms | | | |
|---|---|---|---|---|---|---|---|---|
| Row | $y$ | $z$ | $y'$ | $z'$ | $y+z$ | $y+z'$ | $y'+z$ | $y'+z'$ |
| | | | | | $M_0$ | $M_1$ | $M_2$ | $M_3$ |
| 0 | 0 | 0 | 1 | 1 | 0 | 1 | 1 | 1 |
| 1 | 0 | 1 | 1 | 0 | 1 | 0 | 1 | 1 |
| 2 | 1 | 0 | 0 | 1 | 1 | 1 | 0 | 1 |
| 3 | 1 | 1 | 0 | 0 | 1 | 1 | 1 | 0 |

*Easier way:* You avoid the algebraic manipulations if, for example, you write any the function $F(x, y) = (x + y)(x + y')(x' + y) = M_0 M_1 M_2$ to a K-map (again see page 23). Three maxterms $M_0 M_1 M_2$ means three 0's in the map in squares 0, 1, 2. Therefore square 3 is a 1. Read out the sum of minterms from each map, which is $m_3$.

A table of values defines the following F(y, z). What is the expression for F(y, z) as a product of maxterms, and as a sum of minterms?

| Row | $y$ | $z$ | $F(y,z)$ |
|---|---|---|---|
| 0 | 0 | 0 | 1 |
| 1 | 0 | 1 | 1 |
| 2 | 1 | 0 | 0 |
| 3 | 1 | 1 | 0 |

$0's$ *in rows* $2,3$   $F(y,z) = M_2 M_3 = (y' + z)(y' + z') = y'y' + y'z' + zy' + zz' = y'$

*or* $1's$ *in rows* $0,1$   $F(y,z) = m_0 + m_1 = y'z' + y'z = y'$

*Emphasis* - A row's maxterm is included if the function equals 0 in the same row. Any term can be converted into canonical form by the use of Axiom 4d ($xx' = 0$).

Use DeMorgan's Theorem to prove that a row's maxterm is the complement of the minterm in the same row.

$M_0 = x + y$    $m_0 = x'y'$    $M_0 = m'_0$

$M_1 = x + y'$    $m_1 = x'y$    $M_1 = m'_1$

$M_2 = x' + y$    $m_2 = xy'$    $M_2 = m'_2$

$M_3 = x' + y'$    $m_3 = xy$    $M_3 = m'_3$

## 3.3.3 Perfect Induction

The method of perfect induction is practical for a switching algebra that has only two values. The method is implemented by substituting all combinations of values for the variables in an equation. If the two sides of the equation are equal for all combinations then the equation is proven to be true. Truth tables provide an orderly means for implementing any perfect induction proof. Create a truth table for each side of the equation, and compare the truth tables row by row. If all rows are equal, then the axiom, theorem, or identity is proved. For example:

*Prove Theorem 3d :*  $x + yz = (x + y)(x + z)$

| $x$ | $y$ | $z$ | $yz$ | $x + yz$ | $x + y$ | $x + z$ | $(x + y)(x + z)$ |
|---|---|---|---|---|---|---|---|
| 0 | 0 | 0 | 0 | 0 | 0 | 0 | 0 |
| 0 | 0 | 1 | 0 | 0 | 0 | 1 | 0 |
| 0 | 1 | 0 | 0 | 0 | 1 | 0 | 0 |
| 0 | 1 | 1 | 1 | 1 | 1 | 1 | 1 |
| 1 | 0 | 0 | 0 | 1 | 1 | 1 | 1 |
| 1 | 0 | 1 | 0 | 1 | 1 | 1 | 1 |
| 1 | 1 | 0 | 0 | 1 | 1 | 1 | 1 |
| 1 | 1 | 1 | 1 | 1 | 1 | 1 | 1 |

*Problems* Use perfect induction to prove theorems.

1  $x+0 = x$       1d  $x+1 = 1$

2  $x+y = y+x$      2d  $xy = yx$

3  $x(y+z) = xy+xz$      3d  $x+yz = (x+y)(x+z)$

4  $x+x' = 1$      4d  $x\,x' = 0$

5  $x+x = x$      5d  $x\,x = x$

6  $(x')' = x$

7  $x+1 = 1$      7d  $x \times 0 = 0$

8  $xy + x'z + yz = xy + x'z$      8d  $(x + y)(x'+ z)(y + z) = (x + y)(x'+ z)$

9  $x + xy = x$      9d.    $x(x + y) = x$

10.  $xy + xy' = x$      10d.   $(x + y)(x + y') = x$

11.  $x + x'y = x + y$      11d.   $x(x'+ y) = xy$

12.  $x + (x'+ y) = 1$      12d    $x(x'y) = 0$

13.  $(x + y)' = x'y'$  and for n variables $(x + y + ... + n)' = x'y'.....n'$

13d. $(xy)' = x' + y'$  and for n variables $(xy.....n)' = x' + y' + ... + n'$

14  $x + (y + z) = (x + y) + z$    14d    $x(yz) = (xy)z$

## 3.4 Enumeration of Switching Functions

If there are n variables in a switching equation, then there are $2^n$ rows in the truth table representing the function. In any row, given a switching equation, the function is 0 or 1 for the combination of variables.

If there are $m=2^n$ rows in a table then there are $2^m = 2^{2^n}$ different *OUTPUT columns* of 0s and 1s *representing all possible functions* of n variables.

***Example*** If n = 1 then there are 2 rows and $2^{2^1} = 2^2 = 4$ possible functions of one variable. The four possible functions are displayed in truth table format, and as equations.

| Row | x | $F_0$ | $F_1$ | $F_2$ | $F_3$ |
|-----|---|-------|-------|-------|-------|
| 0 | 0 | 0 | 0 | 1 | 1 |
| 1 | 1 | 0 | 1 | 0 | 1 |

$F_0 = 0, \quad F_1 = x, \quad F_2 = x', \quad F_3 = 1$

***Example*** If n=2 then there are 4 rows and $2^{2^2} = 2^4 = 16$ possible functions of one variable. One function $F_{15}$ includes all 4 minterms: $F_{15}=F(1,1)=1$. Another function $F_0$ does not include any minterms: $F_0=F(0, 0)=0$. The 14 other functions include 1, 2, or 3 minterms in various combinations.

| $x'y'$ | $x'y$ | $xy'$ | $xy$ | sum of minterms | $F(x, y)$ |
|--------|-------|-------|------|-----------------|-----------|
| 0 | 0 | 0 | $0^1$ | 0 | 0 |
| 0 | 0 | 0 | $1^2$ | $xy$ | $xy$ |
| 0 | 0 | 1 | 0 | $xy'$ | $xy'$ |
| 0 | 0 | 1 | 1 | $xy' + xy$ | $x$ |
| 0 | 1 | 0 | 0 | $x'y$ | $x'y$ |
| 0 | 1 | 0 | 1 | $x'y + xy$ | $y$ |
| 0 | 1 | 1 | 0 | $x'y + xy'$ | $x'y + xy'$ |
| 0 | 1 | 1 | 1 | $x'y + xy' + xy$ | $x + y$ |

[1]*omit minterm if 0* [2]*include minterm if 1*

| $x'y'$ | $x'y$ | $xy'$ | $xy$ | sum of minterms | $F(x,y)$ |
|---|---|---|---|---|---|
| 1 | 0 | 0 | $0^1$ | $x'y'$ | $x'y'$ |
| 1 | 0 | 0 | $1^2$ | $x'y'+xy$ | $x'y'+xy$ |
| 1 | 0 | 1 | 0 | $x'y'+xy'$ | $y'$ |
| 1 | 0 | 1 | 1 | $x'y'+xy'+xy$ | $x+y'$ |
| 1 | 1 | 0 | 0 | $x'y'+x'y$ | $x'$ |
| 1 | 1 | 0 | 1 | $x'y'+x'y+xy$ | $x'+y$ |
| 1 | 1 | 1 | 0 | $x'y'+x'y+xy'$ | $x'+y'$ |
| 1 | 1 | 1 | 1 | $x'y'+x'y+xy'+xy$ | 1 |

[1] *omit minterm if 0*      [2] *include minterm if 1*

The 16 functions of two variables are rearranged by five rows according to the operator and number of variables in the function.

| 0 | 1 | | |
|---|---|---|---|
| $x'$ | $x$ | $y'$ | $y$ |
| $x'y'$ | $x'y$ | $xy'$ | $xy$ |
| $x'+y'$ | $x+y'$ | $x'+y$ | $x+y$ |
| $x'y'+xy$ | $x'y+xy'$ | | |

**Example** If n>2 the number $2^{2^n}$ is amazingly large for small values of n.

$$f(n) = 2^{2^n}$$

$$f(n+1) = 2^{2^{n+1}} = 2^{2^n \times 2} = \left[2^{2^n}\right]^2 = [f(n)]^2$$

| $n$ | $f(n) = 2^{2^n}$ |
|---|---|
| 1 | $2^2 = 4$ |
| 2 | $2^4 = 16$ |
| 2 | $2^8 = 256$ |
| 4 | $2^{16} = 65536$ |
| 5 | $2^{32} = 4,294,967,296$ |
| 6 | $2^{64} = (4,294,967,296)^2$ |

# Review 3

***Axioms for a Switching Algebra*** Huntington proved that six axioms provide a complete basis for Boolean Algebra. This becomes a switching algebra when the variables' values are limited to 0 and 1.

***Principle of Duality*** If a Boolean statement is true then the dual of the statement is true.

***Functions*** Functions are algebraic. $F(x,y,z)$ is a sum of terms such as $F(x, y, z) = xyz + z + yz$, and *each* term is a product of literals

Or, a function $F(x,y,z)$ is a product of terms such as $F(x, y, z) = (x + y + z)(z)(y + z)$, and *each* term is sum of literals.

Or, a function $F(x,y,z)$ is a combination of the above. $F(x, y, z) = xz + y + (x + y + z)(z)(y + z)$

***Literal*** A variable x may appear in expressions as x or its complement x'. The term literal is introduced to simplify discussions.

***Minterms*** A minterm is a *canonical product term*. The function $F(x, y)$ which is the sum of one or more minterms, is in a sum of *products canonical form*. For example: $F(x, y) = m_1 + m_2 + m_3 = x'y + xy' + xy$

***Maxterms*** A maxterm is a *canonical sum term*. The function $F(x,y)$ which is the product of one or more maxterms, is in a *product of sums canonical form*. For example: $F(x, y) = M_1 M_2 M_3 = (x + y')(x' + y)(x' + y')$

***M=m'*** Use DeMorgan's Theorem to prove that a row's maxterm is the complement of the minterm in the same row.

***Perfect Induction*** The method of perfect induction is practical for a switching algebra that has only two values. The method is implemented by substituting all combinations of values for the variables in an equation. If the two sides of the equation are equal for all combinations then the equation is proven to be true. This method is not practical for an equation using the real number system with its infinite set of values.

# 4 Building Blocks with no memory

Building blocks with no memory are also known as combinatorial circuits. Their distinctive characteristic is that output signal values are computed from *present* input signal values. A change of any input value produces new outputs after a *time delay* required to compute new values. This *propagation delay* depends upon the implementing technology. Building blocks are commercially available as standard family libraries, and in custom chip libraries.

Digital signals are either at H(igh) voltage or L(ow) voltage. We show how assigning 1 and 0, in different *mixed logic* combinations, to the H and L states of nodes defines $2^{2+1}$ functions for a 2-input, 1-output circuit. Mixed logic recognizes that a node voltage can be assigned true when H *or* L at real circuit nodes. Mixed logic simplifies circuit analysis and design.

We start with the basic building blocks NAND, NOR, NOT, XOR. Next we present the 2 to 1, 4 to 1, and 8 to 1 multiplexers, whose everyday name is mux. We show how to interpret the mux defining equations and related truth tables to create a state machine. We explain the defining equations and related truth tables for the 2 to 4 and 3 to 8 line decoders. Binary arithmetic is implemented with xor gates that implement the add and carry functions of addition as full and half adders. A universal logic circuit (ULC) is an example of a programmable circuit used in a central processing unit (cpu). The ULC is implemented as an arithmetic logic unit. The Manchester full adder is an example of a practical many bit adder..

Boolean equations represent *steady state* behavior. Boolean equations per se do not guarantee correct transient response. We show how *hazards and glitches* arise, and what to do about them, by using *timing diagrams*.

---

*Wires and logic* The following thought contradicts the reality a zero resistance wire is one node. Nevertheless, please tolerate this thought. We will show that logic circuit analysis is facilitated if the two ends of a wire are considered to be two independent nodes, where one node is a gate output and the other node is a gate input. This viewpoint allows us to compare active logic levels at the two end nodes of a wire.

---

# 4.1 Logic Circuit Topics

A two level voltage system is designed to represent switching algebra values 0 and 1. The active level values 0 and 1 assigned to voltage levels can be made in two ways that results in three logic systems known as positive, negative, and mixed (positive and negative). Logic circuits are designed to work in two level voltage systems.

*Active Voltage Levels* When a voltage level, be it H or L, is assigned the value 1 *representing true*, that level is said to be the active level. The other level is assigned the value 0 *representing false*. Voltages are measured at nodes. Consequently each node voltage (representing a variable) is active high or active low. At active high nodes H=1, and at active low nodes L=1. Active low nodes are marked with a *bubble* to distinguish them. Also see Wires and Logic page 62.

| Level | active high nodes | active low nodes |
|-------|-------------------|------------------|
| H     | 1                 | 0                |
| L     | 0                 | 1                |

- *Positive logic* A logic design which includes only active high circuit nodes.
- *Negative logic* A logic design which includes only active low circuit nodes.
- *Mixed logic* A logic design which includes both active high and active low circuit nodes.

*Asserted* An asserted variable is in the true state. An asserted variable is active. An active variable has value 1 that may be either H or L according to the assigned active level. Discussions are simplified when the assertion is what matters, not the active level. When this signal is asserted the bell rings.

*Schematics* Schematics are usually arranged in a hierarchy of levels. For example, the highest level is an interconnected set of circuit building blocks and glue gates. (Glue refers to the few miscellaneous gates that may be required to logically interconnect blocks.) The next level is the set of block schematics. The lowest level is the set of gate schematics showing individual transistors.

## Wires and Logic

We intend to show here that logic circuit analysis is facilitated if the two ends of a wire are considered to be two independent nodes, where one node is a gate output and the other node is a gate input. This viewpoint allows us to compare active logic levels at the two nodes of a wire.

A wire from the active low inverter output z-bubble to a $gate_2$ input connects an active low node to an active high node. We call this an LH wire. Input z to $gate_2$ is active low, whereas input $a_0$ is active high. The $a_0$ wire is an HH wire. Two other HH wires are the wires from active high z and $a_1$ connected to $gate_1$ active high inputs. Observations such as this gave rise to the idea to consider two ends of a wire as two independent nodes.

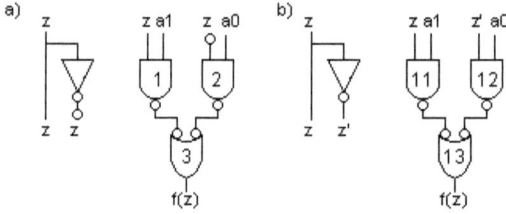

The output of $gate_2$ goes L when *both* inputs z and $a_0$ are H. If $a_0$ is H, then z must be H to drive the $gate_2$ output to L. In other words, z must be false in order to be H.

To show a z false equivalent, we remove the bubble marking z as active low, and *replace z bubble with z tick* (z') as shown at $gate_{12}$.

$gate_2$ *out* = $gate_{12}$ *out* = $z'a_0$

The LL wires from $gate_1$ and $gate_2$ outputs to the $gate_3$ inputs do not create ticks. HH and LL wires do not create ticks (complements).

---

*conclusion*
Tick variables on LH or HL wires. Do not tick variables on LL or HH wires

---

# 4.2 NAND, NOR, XOR

Standard circuit inputs are active high, because nmos transistors (the mn in Figure 401) and npn transistors (not shown here) are turned on by positive input voltage. AND and OR outputs are active low in one stage gates, because a voltage level inversion is a consequence of implementation (as in Figure 401). These are mixed logic circuits.

An nmos transistor, such as $mn_1$ or $mn_2$ (Figure 401), is turned on when the input voltage is at H. Consequently when inputs y & z are H, $mn_1$ and $mn_2$ are turned on discharging node 25 to $V_{SS}$ volts (L). The AND output is active low. A pmos transistor, such as $mp_1$ or $mp_2$, is turned on when its input voltage is at L. So when input z is L $mn_1$ is turned off, and $mp_1$ is turned on charging node 25 to $V_{DD}$ volts (H). This is a mixed logic AND gate, because the input nodes are active high and the output node is active low. By the way, there are other, very different, AND and OR circuits.

These AND and OR logic functions become NAND and NOR functions when a circuit's output is assigned to be active high. The assignment complements the logic function f. To understand this, know that a change in active level from L to H, or H to L, is a logical complement. The complement of 0 is 1 (the complement of L is H), and vice versa. By DeMorgan's Theorem (13d page 45) f=yz (AND) becomes active high output g=(f)'=(yz)'=y'+z' (NAND). Now all nodes are active high. The same happens with the active low output OR.

*In a positive logic system active high gate outputs are connected to the active high inputs of other gates, and that the other gates' active high outputs are in turn connected to yet other gates' active high inputs. And so forth...*

> *The circuit a gate is embedded in defines the gate's logic function.*

**Figure 401 CMOS mixed logic circuits**

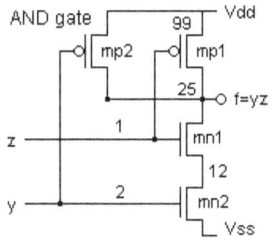
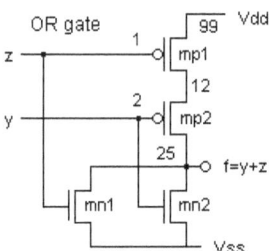

# Digital Design

## 4.2.1 NAND

A gate can have different logic functions f in mixed logic systems. For example, a 2-input NAND[1] gate has 3 terminals (nodes) that can be assigned as active H or L nodes in 8 ways ($2^{2+1}$). We give the 2-input NAND functions the name $f_{nandx}$ where x ranges from 0 to 7. Here only assignments 0, 1, 6, 7 are discussed. (*The standard NAND circuit is 00*)

The 2 input NAND gate physical voltage table shows the relationship of the voltages at the three NAND nodes (inputs y, z, and output $f_{nand}$). Observe the effect of *assigned active node levels* HHL ($f_{nand6}$) and HHH ($f_{nand7}$) on the gate's logic function. The symbol $f_{nand6}$ represents the NAND gate's function with *assigned* active node levels yzf=HHL=$110_2$=$6_{10}$. The $f_{nand6}$ function is AND. The $f_{nand7}$ function is NAND.

| *Physical* | | | *assign* 6 : *yzf* = *HHL* | | | | *assign* 7 : *yzf* = *HHH* | | | |
|---|---|---|---|---|---|---|---|---|---|---|
| y | z | $f_{nand}$ | *min* | y | z | $f_{nand6}$ | *min* | y | z | $f_{nand7}$ |
| L | L | H | $m_0$ | 0 | 0 | 0 | $m_0$ | 0 | 0 | 1 |
| L | H | H | $m_1$ | 0 | 1 | 0 | $m_1$ | 0 | 1 | 1 |
| H | L | H | $m_2$ | 1 | 0 | 0 | $m_2$ | 1 | 0 | 1 |
| H | H | L | $m_3$ | 1 | 1 | 1 | $m_3$ | 1 | 1 | 0 |

It makes sense to give the NAND gate the AND shape (Figure 402a), because inputs y, z must be H to make the output L (see physical table).

**Figure 402a The NAND gate: active levels HHL and HHH (codes 6 and 7)**

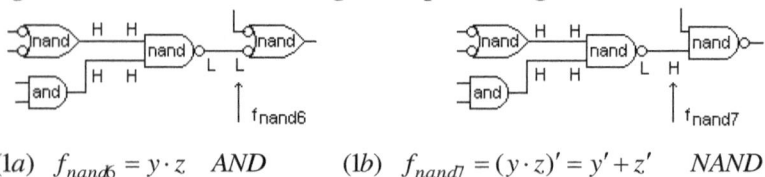

| The circuit a gate is embedded in defines the gate's logic function. |
|---|

**Figure 402b Embedded NAND gates implementing Codes 6 and 7**

(1a) $f_{nand6} = y \cdot z$   AND   (1b) $f_{nand7} = (y \cdot z)' = y' + z'$   NAND

---

[1] The names of the gates (Figure 401) were selected by industry as NAND and NOR, because positive logic at all nodes was assumed.

64

The code 6 active low (L to L) line does not create a tick (see sidebar Wires and Logic). On the other hand a mixed line (H to L or L to H) requires complementing the output function driving the line (to create ticks), because changing the active level means changing a 0 to a 1 or vice versa. E.g. the $f_{nand7}$ output line is an LH mixed logic line (Figure 402b).

The assigned active node levels LLL ($f_{nand0}$) and LLH ($f_{nand1}$) change the NAND gate's logic function to NOR and OR. Note that minterm numbers on the rows change with assignment.

> Minterm numbers are determined by the yz 0,1 values.

| Physical | | | assign 0: $yzf = LLL$ | | | | assign 1: $yzf = LLH$ | | | |
|---|---|---|---|---|---|---|---|---|---|---|
| $y$ | $z$ | $f_{nand}$ | min | $y$ | $z$ | $f_{nand0}$ | min | $y$ | $z$ | $f_{nand1}$ |
| L | L | H | $m_3$ | 1 | 1 | 0 | $m_3$ | 1 | 1 | 1 |
| L | H | H | $m_2$ | 1 | 0 | 0 | $m_2$ | 1 | 0 | 1 |
| H | L | H | $m_1$ | 0 | 1 | 0 | $m_1$ | 0 | 1 | 1 |
| H | H | L | $m_0$ | 0 | 0 | 1 | $m_0$ | 0 | 0 | 0 |

When we focus on the physical voltage table's L level inputs as active, the OR shape (Figure 403a) makes sense for the NAND gate, because when *either* the y or z input is L the output is H. We add input bubbles to the OR shape to represent the yz active L inputs.

**Figure 403a The NAND gate: active levels LLL and LLH (codes 0 and 1)**

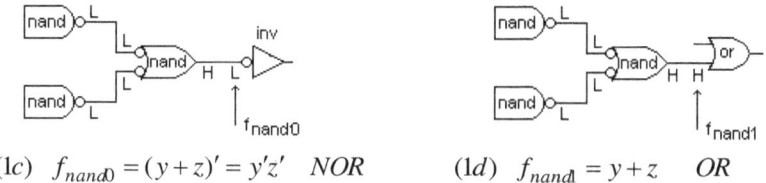

**Figure 403b Embedded NAND gates implementing Codes 0 and 1**

(1c) $f_{nand0} = (y+z)' = y'z'$   NOR   (1d) $f_{nand1} = y+z$   OR

---

*Problem* 401 Sum the minterms to prove $f_{nand6}=yz$.
*Problem* 402 Sum the minterms to prove $f_{nand7}=(yz)'$.
*Problem* 403 Sum the minterms to prove $f_{nand0}=(y+z)'$.
*Problem* 404 Sum the minterms to prove $f_{nand1}=y+z$.

---

## 4.2.2 NOR

The 2-input NOR gate has 3 terminals (nodes) that can be assigned as active H or L nodes in 8 ways. We give the 2-input NOR functions the name $f_{norx}$ where x ranges from 0 to 7. Only assignments 0, 1, 6, 7 are discussed. (*The standard NOR circuit is 02*)

The NOR gate's physical voltage table shows the relationship of the voltages at the three NOR nodes (inputs y, z, and output $f_{nor}$). Observe the effect of *assigned active node levels* HHL ($f_{nor6}$) and HHH ($f_{nor7}$) on the gate's logic function. The symbol $f_{nor6}$ is the NOR gate's function with *assigned* active node levels yzf=HHL=$110_2$=$6_{10}$. The $f_{nor6}$ function is OR. The $f_{nor7}$ function is NOR.

| Physical | | | assign 6 : yzf = HHL | | | | assign 7 : yzf = HHH | | | |
|---|---|---|---|---|---|---|---|---|---|---|
| y | z | $f_{nor}$ | min | y | z | $f_{nor6}$ | min | y | z | $f_{nor7}$ |
| L | L | H | $m_0$ | 0 | 0 | 0 | $m_0$ | 0 | 0 | 1 |
| L | H | L | $m_1$ | 0 | 1 | 1 | $m_1$ | 0 | 1 | 0 |
| H | L | L | $m_2$ | 1 | 0 | 1 | $m_2$ | 1 | 0 | 0 |
| H | H | L | $m_3$ | 1 | 1 | 1 | $m_3$ | 1 | 1 | 0 |

If we focus on the physical voltage table's H level inputs as active, the OR shape (Figure 404a) makes sense, because when *either* the y or z input is H the output is L.

**Figure 404a The NOR gate: active levels HHL and HHH (codes 6 and 7)**

> The circuit a gate is embedded in defines the gate's logic function.

**Figure 404b Embedded NOR gates implementing Codes 6 and 7**

(2a)  $f_{nor6} = y + z$   OR          (2b)  $f_{nor7} = (y + z)' = y'z'$   NOR

The code 6 active low (L to L) line does not create a tick (see sidebar Wires and Logic). On the other hand the code 7 mixed line (L to H) creates a tick.

The assigned active node levels LLL ($f_{nor0}$) and LLH ($f_{nor1}$) change the NOR gate's logic function to NAND and AND. Note that minterm numbers on the rows change with assignment.

| Minterm numbers are determined by the yz 0,1 values. |
| --- |

| Physical | | | assign 0: $yzf = LLL$ | | | | assign 1: $yzf = LLH$ | | | |
| --- | --- | --- | --- | --- | --- | --- | --- | --- | --- | --- |
| $y$ | $z$ | $f_{nor}$ | min | $y$ | $z$ | $f_{nor0}$ | min | $y$ | $z$ | $f_{nor1}$ |
| L | L | H | $m_3$ | 1 | 1 | 0 | $m_3$ | 1 | 1 | 1 |
| L | H | L | $m_2$ | 1 | 0 | 1 | $m_2$ | 1 | 0 | 0 |
| H | L | L | $m_1$ | 0 | 1 | 1 | $m_1$ | 0 | 1 | 0 |
| H | H | L | $m_0$ | 0 | 0 | 1 | $m_0$ | 0 | 0 | 0 |

When we focus on the physical voltage table's L level inputs as active, the AND shape (Figure 405a) makes sense for the NOR gate, because when *both* inputs yz are L the output is H. We add input bubbles to the AND shape to represent the L active level inputs.

**Figure 405a The NOR gate: active levels LLL and LLH (codes 0 and 1)**

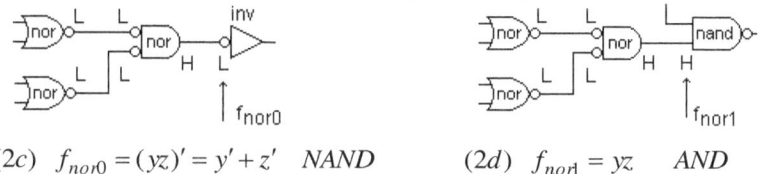

**Figure 405b Embedded NOR gates implementing Codes 0 and 1**

(2c) $f_{nor0} = (yz)' = y' + z'$    *NAND*        (2d) $f_{nor1} = yz$    *AND*

---

*Problem* 405 Sum the minterms to prove $f_{nor7}=(y+z)'$.
*Problem* 406 Sum the minterms to prove $f_{nor6}=y+z$.
*Problem* 407 Sum the minterms to prove $f_{nor1}=yz$.
*Problem* 408 Sum the minterms to prove $f_{nor0}=(yz)'$.

---

## 4.2.3 XOR

The 2 input XOR gate has two inputs and one output. The XOR gate's voltage table shows the relationship of the voltages at the three terminals of the XOR that has 8 active level assignments (codes 0 to 7). The XOR uses the unique *exclusive OR* shape (Figure 406a), and operator $\oplus$. (*The standard XOR circuit is 86*)

The XOR output is low only when inputs are the same.
The XOR output is high when the inputs are different.

**Figure 406a The XOR has 8 active level combinations (code yzf: 0 to 7)**

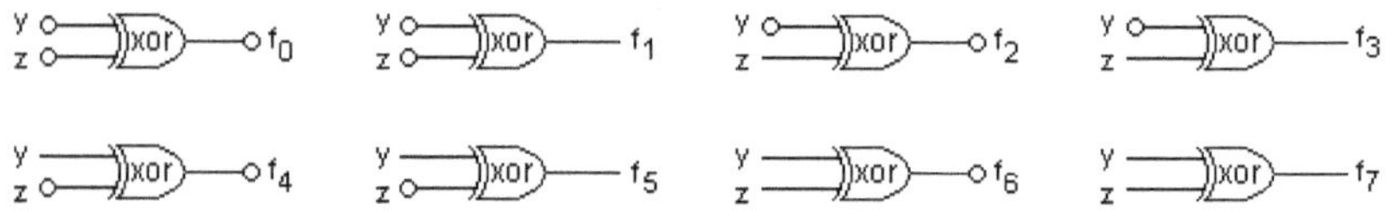

Observe the effect of active level assignments HHL ($f_{xor6}$) and HHH ($f_{xor7}$) on the gate's logic function.

| $y$ | $z$ | $f_{xor}$ | Assign<br>min | H<br>$y$ | H<br>$z$ | L<br>$f_{xor6}$ | H<br>$f_{xor7}$ |
|---|---|---|---|---|---|---|---|
| $L$ | $L$ | $L$ | $m_0$ | 0 | 0 | 1 | 0 |
| $L$ | $H$ | $H$ | $m_1$ | 0 | 1 | 0 | 1 |
| $H$ | $L$ | $H$ | $m_2$ | 1 | 0 | 0 | 1 |
| $H$ | $H$ | $L$ | $m_3$ | 1 | 1 | 1 | 0 |

> *The circuit a gate is embedded in defines the gate's logic function.*

**Figure 406b Embedded XOR gates implementing Codes 6 and 7**

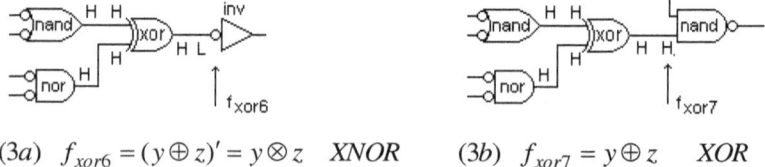

(3a) $f_{xor6} = (y \oplus z)' = y \otimes z$ XNOR $\quad$ (3b) $f_{xor7} = y \oplus z$ XOR

Now try active level assignments LLL ($f_{xor0}$) and LLH ($f_{xor1}$).

| | | | *Assign* | L | L | L | H |
|---|---|---|---|---|---|---|---|
| $y$ | $z$ | $f_{xor}$ | min | $y$ | $z$ | $f_{xor0}$ | $f_{xor1}$ |
| L | L | L | $m_3$ | 1 | 1 | 1 | 0 |
| L | H | H | $m_2$ | 1 | 0 | 0 | 1 |
| H | L | H | $m_1$ | 0 | 1 | 0 | 1 |
| H | H | L | $m_0$ | 0 | 0 | 1 | 0 |

**Figure 406c Embedded XOR gates implementing Codes 0 and 1**

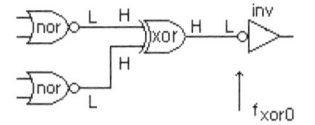

(3c)  $f_{xor0} = (y' \oplus z')' = y \otimes z$   *XNOR*    (3d)  $f_{xor1} = y' \oplus z' = y \oplus z$   *XOR*

Next we try something different: codes 2 and 5. Observe that the truth table *y,z rows* are *not* sequential.

| | | | | | | | | | | | |
|---|---|---|---|---|---|---|---|---|---|---|---|
| *Physical* | | | | *assign 2 LHL* | | | | *assign 5 HLH* | | | |
| $y$ | $z$ | $f_{xor}$ | | min | $y$ | $z$ | $f_{xor2}$ | min | $y$ | $z$ | $f_{xor5}$ |
| L | L | L | | $m_2$ | 1 | 0 | 1 | $m_1$ | 0 | 1 | 0 |
| L | H | H | | $m_3$ | 1 | 1 | 0 | $m_0$ | 0 | 0 | 1 |
| H | L | H | | $m_0$ | 0 | 0 | 0 | $m_3$ | 1 | 1 | 1 |
| H | H | L | | $m_1$ | 0 | 1 | 1 | $m_2$ | 1 | 0 | 0 |

> *Write the minterms in numerical sequence to see the function defined.*

**Figure 406d Embedded XOR gates implementing Codes 2 and 5**

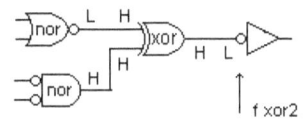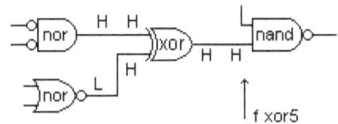

(3e)  $f_{xor2} = (y' \oplus z)' = x \oplus y$   *XOR*    (3f)  $f_{xor5} = y \oplus z' = y \otimes z$   *XNOR*

---

*Problem* 409 Sum the minterms to prove $f_{xor6}=(y \text{ xor } z)'$.
*Problem* 410 Sum the minterms to prove $f_{xor7}=y \text{ xor } z$.
*Problem* 411 Sum the minterms to prove $f_{xor2}=y \text{ xor } z$.
*Problem* 412 Sum the minterms to prove $f_{xor5}=(y \text{ xor } z)'$.

---

## Timing diagrams

A *timing diagram* is a graphic illustrating the behavior of a circuit's logic as a function of time. A timing diagram's horizontal axis represents time. The vertical axis usually represents analog voltage or current. A *digital waveform* is a plot of the High and Low levels a node variable takes as a function of time (Figure 400). Transitions from L to H require a finite time called *rise time*. An H to L transition duration is called *fall time*. Ideal waveforms with zero rise and fall times (Figure 400b) are used to simplify the graphic. Finite rise and fall times are usually shown when limits of performance are evaluated. A clock's periodic waveform (Figure 400c) shows waveform parameters such as rising edge, falling edge, and period.

Each digital waveform represents logical behavior at a circuit node. The waveforms at a set of circuit nodes are related. These are cause and effect relationships. Cause and effect are shown (Figure 400d) by arrows from the transitions in one waveform (*cause*) to the transitions in other waveforms (*effect*). Cause produces effect after a delay. Industrial practice measures delays from the cause's 50% level to the effect's 50% level. The delay arises from a signal traveling through gate circuits and along signal lines at the speed of light in the medium such as 30cm/ns divided by the square root of the dielectric constant of the medium.

A timing diagram is related to a *timing table* specifying symbols for delays, values for time delays. and the conditions under which they apply. Timing tables are included in specifications found in data books. Tables show minimum, typical, and maximum values for each timing parameter. The range of possible delays arises from variations in manufacturing processes, circuit supply voltage, and temperature.

**Figure 400 Timing Diagrams**

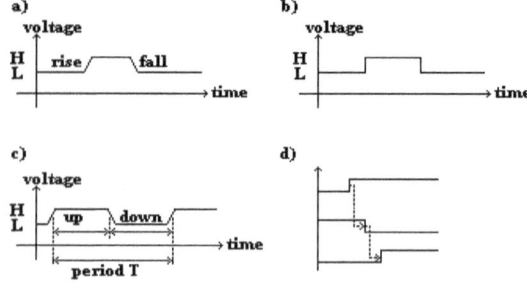

# 4.3 Multiplexers

Multiplexing is a method by which N address lines x, y, z ... select one out of a set of $2^N$ input signals $a_K$. The selected $a_K$ signal is routed to the multiplexer (mux) output. For example, if N=1 there are two input signals $a_1$, $a_0$, and one address line z is required for selection. The multiplexer output equation is $f(z)=a_0 z'+a_1 z$. When z=L, $z_{bubble}=z'=H$, and input $a_0$ is selected (Figure 407). When z=H, $z_{bubble}=z'=L$, and input $a_1$ is selected. Each term is the output of one AND.

**Figure 407 Two to one multiplexer - two equivalent schematics**

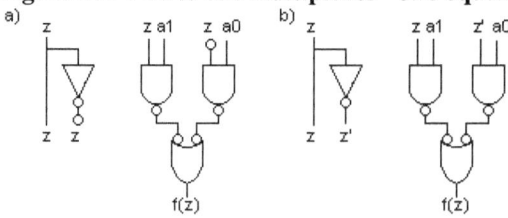

The NOT gate produces an active low z line at the bubble output (Figure 407a) that is also referred to as an active high z' line (Figure 407b). If address line z is H, then $f(z)=a_0 z'+a_1 z = a_0 L+a_1 H= a_0 0+a_1 1=a_1$. If z is L, then $f(z)=a_0 z'+a_1 z = a_0 H+a_1 L= a_0 1+a_1 0=a_0$.

*Emphasis: active low z-bubble is an active high z'.*

Observe that the lines from the AND outputs to the OR inputs are active low at both ends, and so they are not in conflict. No ticks are required.

> The 3 gates are NAND gates, two if which behave as AND ($f_{nand6}$) & one behaves as OR ($f_{nand1}$). This is a mixed logic design example.

In another example assume there are four input signals $a_3$, $a_2$, $a_1$, $a_0$. Since $4=2^2=2^N$ N=2 and two address lines y, z are required for selection. The multiplexer output equation is $f(y,z) = yz a_3 + yz'a_2 + y'z a_1 + y'z'a_0$, where each term is the output of one AND. This 4 input to 1 output mux requires one 3 input AND for each input signal, one 4 input OR to route the selected input from its corresponding AND output to the mux output, and two NOTs forming complements of two address lines (Figure 408).

# Digital Design

A mux is an OR of ANDs logic circuit, because AND represents selection of one input, and OR has the ability to route any selected input to the output. NOT gates are used to form the complements of address variables that are used to select an input. (This is an example of *decoding* an address such as yz.)

**Figure 408 Four to one mux**

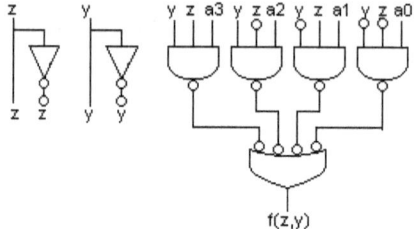

f(z,y)

***How the mux works***: In industrial practice mux data inputs and output are active high variables. Two address lines represent two binary digits (bits). Two bits can represent four numbers 0, 1, 2, and 3. The $a_K$ input signals are table entered variables

| *min* | y | z | y'z' | y'x | yz' | yz |
|-----|---|---|------|-----|-----|-----|
| $m_0$ | 0 | 0 | $a_0$ | 0 | 0 | 0 |
| $m_1$ | 0 | 1 | 0 | $a_1$ | 0 | 0 |
| $m_2$ | 1 | 0 | 0 | 0 | $a_2$ | 0 |
| $m_3$ | 1 | 1 | 0 | 0 | 0 | $a_3$ |

The NOT circuits produce active high complements y' and z'. Each 3 input AND has one $a_k$ input, one y or y' input, and one z or z' input. For example, row 2 requires y, z values of 1, 0 so that y, z' have values 1, 1 so that the AND with inputs yz' (minterm $m_2$) is asserted.

*Address k selects minterm k*

Note: Data selector is another name for multiplexer.

A multiplexer usually has one or more enable inputs g that activate the circuit.

## 4.3.1 The 2:1, 4:1, and 8:1 Multiplexers

**The 2:1 mux** (*The standard 2:1 mux circuit is 157*) The mux address line $z$ implements two addresses 0 and 1. Address $z$ selects one of two signal inputs $a_0$ or $a_1$ (Figure 409). Observe how the output f changes from $a_0$ to $a_1$ to $a_0$ when $z$ changes from L to H to L. Z is shown as a periodic waveform (*which z does not have to be*). The L input at g enables the mux.

**Figure 409 2 line to 1 line multiplexer**

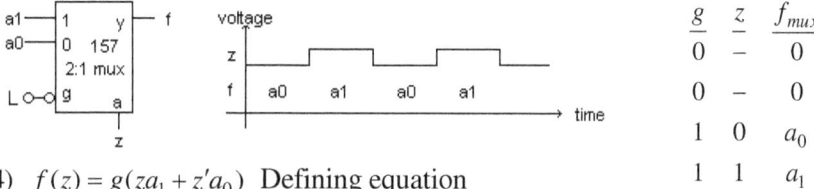

| g | z | $f_{mux}$ |
|---|---|---|
| 0 | – | 0 |
| 0 | – | 0 |
| 1 | 0 | $a_0$ |
| 1 | 1 | $a_1$ |

(4)  $f(z) = g(za_1 + z'a_0)$  Defining equation
(5)  $f = g(m_1a_1 + m_0a_0)$  Minterm format

**The 4:1 mux** (*The standard 4:1 mux circuit is 153*) The mux address lines $y$, $z$ implement four addresses 00, 01, 10, and 11 (decimal 0, 1, 2, 3). Each address selects one of four signal inputs $a_0$ to $a_3$ (Figures 408, 410) when the enable input g=L=1.

**Figure 410 4 line to 1 line multiplexer**

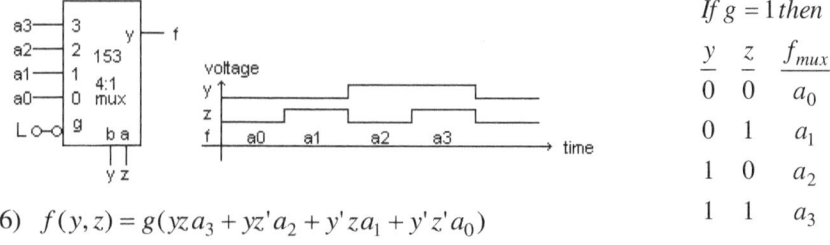

If $g = 1$ then

| y | z | $f_{mux}$ |
|---|---|---|
| 0 | 0 | $a_0$ |
| 0 | 1 | $a_1$ |
| 1 | 0 | $a_2$ |
| 1 | 1 | $a_3$ |

(6)  $f(y,z) = g(yza_3 + yz'a_2 + y'za_1 + y'z'a_0)$

---

*Problems* 413-417 Use one 4:1 mux and gates, if necessary, to synthesize the following functions.
413 f = (x xor y)z
414 f = [xy + (x xor y)]z
415 f = [xy + z(x + y)]
416 f = x+ y'z
417 f = x+ z' + y'

---

Digital Design

**The 8:1 mux** (*The standard 8:1 mux circuit is 151*) The mux address lines
$x, y, z$ implement eight addresses 000, 001, 010, 011, 100, 101, 110, and
111 (decimal 0 to 7). Each address selects one of eight signal inputs $a_0$ to
$a_7$ (Figure 411) when the enable input g=1.

**Figure 411 8 line to 1 line multiplexer**

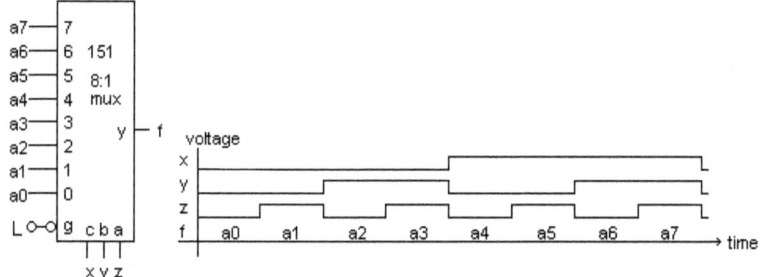

$$(7)\ f(x, y, z) = g(xyz\,a_7 + xyz'a_6 + xy'za_5 + xy'z'a_4 + x'yza_3 + x'yz'a_2 + x'y'za_1 + x'y'z'a_0)$$

*If g = 1 then*

| $x$ | $y$ | $z$ | $f_{mux}$ |
|---|---|---|---|
| 0 | 0 | 0 | $a_0$ |
| 0 | 0 | 1 | $a_1$ |
| 0 | 1 | 0 | $a_2$ |
| 0 | 1 | 1 | $a_3$ |
| 1 | 0 | 0 | $a_4$ |
| 1 | 0 | 1 | $a_5$ |
| 1 | 1 | 0 | $a_6$ |
| 1 | 1 | 1 | $a_7$ |

---

*Problems* 418-422 Use one 8:1 mux and gates, if necessary, to synthesize
the following functions.
418 f = x xor y xor z
419 f = xy + z(x xor y)
420 f = xy + z(x + y)
421 f = wx+ y'z
422 f = wx+ wz' + wy' + yz'

---

74

## 4.3.2 Multiplexer Application: ASM

The d flip-flop input equations developed in Top Down Design page 10 were of a random logic nature. Here we convert them to mux format by expanding the equations to include states with coefficient 0. The coefficients are the mux inputs.

$(8a)\quad d_2 = S_3 \times kbd + S_4 \times kbd'$

$\qquad = 0 \times S_0 + 0 \times S_1 + 0 \times S_2 + kbd \times S_3 + kbd' \times S_4$

$(8b)\quad d_1 = S_1 + S_3 \times kbd'$

$\qquad = 0 \times S_0 + 1 \times S_1 + 0 \times S_2 + kbd' \times S_3$

$(8c)\quad d_0 = S_0(R<N)(R=0)' + S_3 \cdot kbd'$

$\qquad = (R<N) \times (R=0)' \times S_0 + 0 \times S_1 + 0 \times S_2 + kbd \times S_3$

**Figure 412 ASM implemented with Multiplexers**

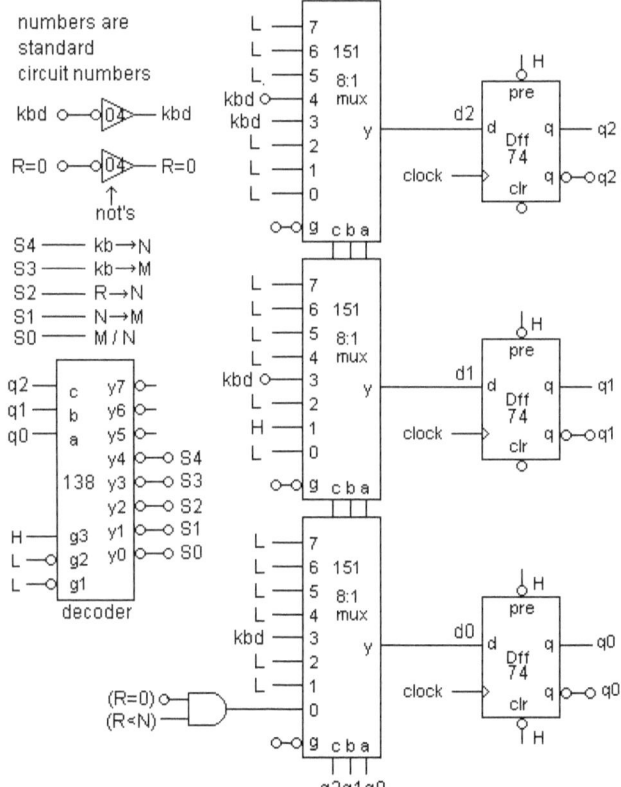

# 4.4 Decoders

A decoder creates minterms. For example, a 3 to 8 decoder converts a 3 digit binary number into a minterm. Binary input 011 produces output $m_3$. In other words the purpose of a decoder is to assert one output node for each minterm of the input variables. Decoder input variables are referred to as *address lines*. N address lines represent $2^N$ different values ranging from 0 to $2^N-1$. For example $y_7 = a_3'a_2a_1a_0$ equals 1 when the input variables are $0111_2$. Output $y_7$ represents the number $7_{10}$ (minterm $m_7$). The four address lines implement $2^4$ or 16 addresses.

A complex example is a memory chip. If the chip stores 1024 words (a word is a group of bits such as 32 bits), then the chip binary decoder requires ten address lines because $1024=2^{10}$. The memory decoder (Chapter 8) in effect consists of 1024 ten input AND gates, so that words stored at addresses 0 to 1023 can be selected for reading or writing. The ancient analogy is this: think of addressing post office boxes.

A decoder circuit uses a set of independent AND gates, and NOT gates are used to form the complements $a_K'$ of address variables $a_K$ as required. A decoder has enable inputs that activate the decoder. N address bits on address lines assert one of $2^N$ outputs. *Decoders are minterm generators.*

***The 2:4 decoder*** (*The standard 2:4 decoder circuit is 139*) The 2:4 decoder building block is an assembly of four 3 input NAND and 2 NOT gates. Address lines $y$, $z$ connected to decoder inputs a,b assert one of four output nodes $y_0$ to $y_3$ when enable g is true (Figure 413). Note the active low outputs. The defining equations and the truth table are as follows. (The address signals do not have to be periodic.)

(9)  $y_0 = gy'z'$   $y_1 = gy'z$   $y_2 = gyz'$   $y_3 = gyz$

*If $g = 1$ then*

| $y$ | $z$ | $y_0$ | $y_1$ | $y_2$ | $y_3$ |
|---|---|---|---|---|---|
| 0 | 0 | 1 | 0 | 0 | 0 |
| 0 | 1 | 0 | 1 | 0 | 0 |
| 1 | 0 | 0 | 0 | 1 | 0 |
| 1 | 1 | 0 | 0 | 0 | 1 |

**Figure 413 2 line to 4 line decoder**

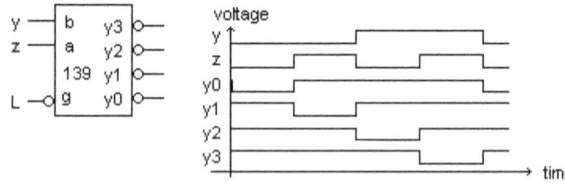

***The* 3:8 *decoder*** (*The standard 3:8 decoder circuit is 138*) The 3 line to 8 line decoder has address lines *x, y, and z* connected to decoder inputs a,b,c that assert one of eight active low output nodes $y_0$ to $y_7$ when enable $g_1 g_2 g_3$ is true (Figure 414). The defining equation shows that the decoder circuit is simply 8 4-input NAND and 3 NOT gates. Note: The address signals do not have to be periodic.

(10) *If* $g = g_1 g_2 g_3$ *then*

$$y_0 = gx'y'z' \quad y_1 = gx'y'z \quad y_2 = gx'yz' \quad y_3 = gx'yz$$
$$y_4 = gxy'z' \quad y_5 = gxy'z \quad y_6 = gxyz' \quad y_7 = gxyz$$

**Figure 414 The 138 3 line to 8 line decoder**

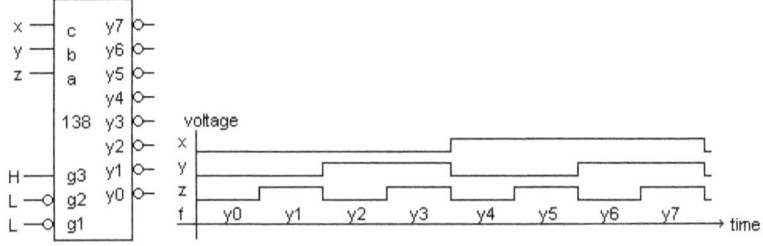

---

*Problems* 423-427 Use a 3:8 decoder and gates, if necessary, to synthesize the following functions.

423 f = (x xor y)z                424 f = [xy + (x xor y)]z

425 f = [xy + z(x + y)]          426 f = x+ y'z                427 f = x+ z' + y'

*Problem* 428 Use a 3:8 decoder and gates to decode an eight bit address $a_K$ (k = 0 to 7) by producing asserted low output lines for hex addresses 87, 8F, 97, 9F, A7, AF, B7, and BF.

*Problem* 429 Use two 3:8 decoders and no gates to synthesize $f_0$ and $f_1$.

$f_0 = a_5 a_4 a_3' a_2' a_1 a_0'$                $f_1 = a_5' a_4' a_3 a_2 a_1' a_0$

*Problems* 430-431 Use multiplexers and supplementary gates to design Gray code to binary code converters defined by the following truth table. (430 Use 3:8 decoders plus gates.) (431 Use 2:4 decoders plus gates.)

*Gray code*   0132 6754 *CDFE AB*98

*binary code*   0123 4567 89*AB CDEF*

Inputs-Gray wxyz        Outputs-Binary $y_3 y_2 y_1 y_0$

---

# 4.5 Binary Arithmetic Logic Circuits

Two is the base (radix) of the binary number system. Base 2 numbers use the two digits 0 and 1 such as in the number $1101_2$ that equals $13_{10}$. The basic calculating functions are implemented by XOR and AND. We start with addition. (*The standard 1 bit pair adder circuit is 86*)

## 4.5.1 Addition

The addition table for one digit numbers is straightforward. The addition table entry 1 plus $1=1+1=10_2$ differs from the logical 1 OR $1=1+1=1$.

$$
\begin{array}{cccc}
0 & 0 & 1 & 1 \\
(11) \quad +0 & +1 & +0 & +1 \\
\hline
00 & 01 & 01 & 10
\end{array}
$$

In decimal arithmetic the carry is 1 when 1 is added to greatest digit 9, and the sum is 0 in the current column (9+1=10, ten). The greatest digit in binary arithmetic is 1 so the carry is one when 1 is added to 1 ($1+1=10_2$, two). In the half adder truth table, the sum function is XOR, and the carry function is AND. A half adder has no carry input. A full adder provides for input carries to the current digit column from the digit column at the right (in many digit numbers). A carry is produced when at least two out of the three inputs x, y, $c_{in}$ equal 1.

*Full adder*

$$s = x \oplus y \oplus c_{in}$$
$$c = xy + c_{in}(x+y)$$

*Half adder* $s = x \oplus y$ *and* $c = xy$

| Row | x | y | sum s | carry c |
|-----|---|---|-------|---------|
| 0 | 0 | 0 | 0 | 0 |
| 1 | 0 | 1 | 1 | 0 |
| 2 | 1 | 0 | 1 | 0 |
| 3 | 1 | 1 | 0 | 1 |

| Row | $c_{in}$ | x | y | s | c |
|-----|----------|---|---|---|---|
| 0 | 0 | 0 | 0 | 0 | 0 |
| 1 | 0 | 0 | 1 | 1 | 0 |
| 2 | 0 | 1 | 0 | 1 | 0 |
| 3 | 0 | 1 | 1 | 0 | 1 |
| 4 | 1 | 0 | 0 | 1 | 0 |
| 5 | 1 | 0 | 1 | 0 | 1 |
| 6 | 1 | 1 | 0 | 0 | 1 |
| 7 | 1 | 1 | 1 | 1 | 1 |

The addition of larger numbers is a straightforward extension of how one digit numbers are added. Add *same-weight-digits* and carry to the left. Same-weight-digits in binary numbers have weights 1. 2, 4, 8, 16, etc. Consider the equivalent binary sum of the decimal sum 13+11=24.

| | |
|---|---|
| *4 3 2 1 0* | $\leftarrow$ *digit position* |
| *1 1 1 1 0* | $\leftarrow$ *carries* |
| *1 1 0 1* | $\leftarrow 13_{10}$ |
| *+1 0 1 1* | $\leftarrow +11_{10}$ |
| *1 1 0 0 0* | $\leftarrow 24_{10}$ |

***Circuits for adding many digit binary numbers*** The ones column is implemented by a half adder, and full adders implement the digit columns to the left. The important question is *When does the correct answer appear at the outputs?* Assume the bits of both numbers are presented to the adder inputs simultaneously. Furthermore assume $t_D$ is the time to add two one digit numbers, and $t_D$ is also the time to produce a carry at the carry input of the column to the left. The *ripple carry* and *look ahead carry* circuits are the two classic circuits.

***Ripple carry*** At time t=0 the sum and carry outputs have no meaning. At time $t_D$ sum and carry values appear at all of the outputs (Figure 419). Except for the one's column, the outputs are not correct! They are not correct, because carry inputs to more significant digits have not had time to appear. At time $2t_D$ the two's column's correct $sum_1$ and carry $c_1$ appear. The correct values appear as the carries *ripple* from column to column. Here is how we develop the addition time estimate of $5t_D$ .

| | | | | | |
|---|---|---|---|---|---|
| 1 | 0 | 0 | 1 | 0 | c |
| | 1 | 1 | 0 | 1 | x |
| + | +1 | +0 | +1 | +1 | y |
| 0 | 10 | 01 | 10 | 10 | s | $t = 2t_d$ |

| | | | | | |
|---|---|---|---|---|---|
| 1 | 0 | 1 | 1 | 0 | c |
| | 1 | 1 | 0 | 1 | x |
| + | +1 | +0 | +1 | +1 | y |
| 1 | 10 | 10 | 10 | 10 | s | $t = 3t_d$ |

$$
\begin{array}{ccccc}
1 & 1 & 1 & 1 & 0 \quad c \\
  & 1 & 1 & 0 & 1 \quad x \\
+ & +1 & +0 & +1 & +1 \quad y \\
\hline
1 & 10 & 10 & 10 & 10 \quad s \qquad t = 4t_d
\end{array}
$$

$$
\begin{array}{ccccc}
1 & 1 & 1 & 1 & 0 \quad c \\
  & 1 & 1 & 0 & 1 \quad x \\
+ & +1 & +0 & +1 & +1 \quad y \\
\hline
1 & 11 & 10 & 10 & 10 \quad s \qquad t = 5t_d \\
1 & 1 & 0 & 0 & 0 \qquad\quad sum
\end{array}
$$

***Full Adder with Look ahead carry*** (*The standard 4-bit look ahead carry adder circuit is 283*) Ripple carry schemes are not satisfactory in many applications. Do you want to wait while a ripple carry adder adds 32 bits? No? How to start? We rediscover a solution when we analyze the carry equations. From the truth table for a full adder and the block diagram for a 4 bit adder (Figure 415) we proceed by simplifying the full adder carry out equation (see Full Adder Truth Table).

$$
\begin{aligned}
(12) \quad c_{out} &= xy + c_{in}(x \oplus y) \\
&= xy + c_{in}(x+y)(x'+y') = (xy) + c_{in}(x+y)(xy)' \\
&= xy + c_{in}(x+y)
\end{aligned}
$$

Write the $c_{OUT}$ equation for the four carries.
$(13a) \quad c_0 = x_0 y_0 + c_{in}(x_0 + y_0)$
$(13b) \quad c_1 = x_1 y_1 + c_0(x_1 + y_1)$
$(13c) \quad c_2 = x_2 y_2 + c_1(x_2 + y_2)$
$(13d) \quad c_3 = x_3 y_3 + c_2(x_3 + y_3)$

**Figure 415 Four bit Adder (283) with look ahead carry - sum=a+b**

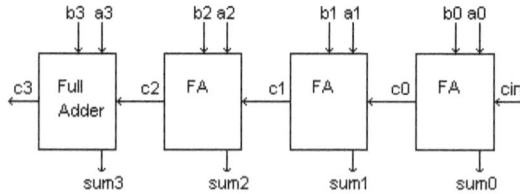

Observe that $c_0$ has to wait for $c_{in}$, $c_1$ has to wait for $c_0$, $c_2$ has to wait for $c_1$, and so forth. Also observe that if $c_0$ is eliminated from the $c_1$ equation $c_1$ does not have to wait for $c_0$. The goal is to only have $c_{in}$ in each carry equation. To simplify the algebra define g and p and note that gp=g.

(14a) $\quad g = xy \text{ and } p = x + y$

(14b) $\quad gp = xy(x+y) = xyx + xyy = xy + xy = xy = g$

The g and p represent *generate* and *propagate* respectively. When two same weight digits both equal 1 then a carry is generated whether an input carry is 0 or 1. If the input carry is one and *either* same-weight-bit is one, then a carry is propagated to the output. For any bit column

(15) $\quad c_{out} = xy + c_{in}(x+y) = g + pc_{in}$ so that

(16a) $\quad c_0 = g_0 + p_0 c_{in}$

(16b) $\quad c_1 = g_1 + p_1 c_0$
$$= g_1 + p_1(g_0 + p_0 c_{in})$$
$$= g_1 + p_1 g_0 + p_1 p_0 c_{in}$$

(16c) $\quad c_2 = g_2 + p_2 c_1$
$$= g_2 + p_2(g_1 + p_1 g_0 + p_1 p_0 c_{in})$$
$$= g_2 + p_2 g_1 + p_2 p_1 g_0 + p_2 p_1 p_0 c_{in}$$

(16d) $\quad c_3 = g_3 + p_3 c_2$
$$= g_3 + p_3(g_2 + p_2 g_1 + p_2 p_1 g_0 + p_2 p_1 p_0 c_{in})$$
$$= g_3 + p_3 g_2 + p_3 p_2 g_1 + p_3 p_2 p_1 g_0 + p_3 p_2 p_1 p_0 c_{in}$$

These are two gate level OR of ANDs equations whose outputs are simultaneously available. Total delay is $t_D$, and the delay is independent of the number of bits to be added. The circuit for $bit_3$ has 4 AND gates and 1 five input OR gate. *Do not even think about the circuit for $bit_{31}$!*

---

*This is why a focus of research is about the many-bit-adder problem.* The Manchester Adder (4.5.3 page 84) is one result of this research.

---

Problem 432 Continue from $c_3$ and write the equation for $c_4$.

Problem 433 Convert $c_1$ from g and p expressions to x and y expressions.

---

## 4.5.2 Universal Logic Circuit

The universal logic circuit (ULC Figure 416) is a circuit whose function is *programmed* by bits $s_j$. This ULC is designed to generate the 16 functions of two variables. The ULC functions are in the 4 bit 181 ALU. (*The standard ALU and carry circuits are 181, 182*)

**Figure 416 Universal Logic Circuit**

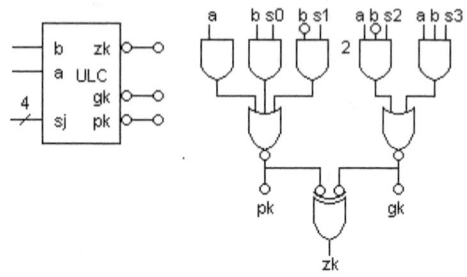

*Defining equations :*

(17a) $z_k = g_k \oplus p_k$

(17b) $g_k = (s_3 ba + s_2 b'a)$

(17c) $p_k = (s_1 b' + s_0 b + a)$

*16 functions of 2 variables*

| 0 | | 1 | |
|---|---|---|---|
| $a'$ | $a$ | $b'$ | $b$ |
| $a'b'$ | $a'b$ | $ab'$ | $ab$ |
| $a'+b'$ | $a'+b$ | $a+b'$ | $a+b$ |
| $a'b'+ab$ | $ab'+a'b$ | | |

(35) $z_k = g_k \oplus p_k = (s_3 ba + s_2 b'a) \oplus (s_1 b' + s_0 b + a)$

| $s_{3210}$ | $z_k = g_k \oplus p_k$ | $z_k$ |
|---|---|---|
| 0000 | $z_0 = 0 \oplus a$ | $= a$ |
| 0001 | $z_1 = 0 \oplus (b+a)$ | $= b+a$ |
| 0010 | $z_2 = 0 \oplus (b'+a)$ | $= b'+a$ |
| 0011 | $z_3 = 0 \oplus (b'+b+a)$ | $= 1$ |
| 0100 | $z_4 = b'a \oplus a$ | $= ba$ |
| 0101 | $z_5 = b'a \oplus (b+a)$ | $= b$ |
| 0110 | $z_6 = b'a \oplus (b'+a)$ | $= b' \oplus a$ |
| 0111 | $z_7 = b'a \oplus (b'+b+a)$ | $= b+a'$ |
| 1000 | $z_8 = ba \oplus a$ | $= b'a$ |
| 1001 | $z_9 = ba \oplus (b+a)$ | $= b \oplus a$ |
| 1010 | $z_A = b'a \oplus (b'+a)$ | $= b'$ |
| 1011 | $z_B = ba \oplus (b'+b+a)$ | $= b'+a'$ |
| 1100 | $z_C = (ba+b'a) \oplus a$ | $= 0$ |
| 1101 | $z_D = (ba+b'a) \oplus (b+a)$ | $= ba'$ |
| 1110 | $z_E = (ba+b'a) \oplus (b'+a)$ | $= b'a'$ |
| 1111 | $z_F = (ba+b'a) \oplus (b'+b+a)$ | $= a'$ |

*Problem* 434 Reference the ULC. Show that $g_k p_k = g_k$.

*Problem* 435 Show that $g_4 p_4 = g_4$.

*Problem* 436 Use the ULC z equation to demonstrate that
$$z_6 = g_6 \oplus p_6 = b' \oplus a \quad or \quad z_6 = b \oplus a'$$

*Problem* 437 Show that the sum of two bits is converted from xor to s=g'p by using the trick of adding x'x+y'y so that
$$\Sigma = s \oplus c = x \oplus y \oplus c = g'p \oplus c$$

*Problem* 438 Show that $z_9 = b$ xor a.

*Problem* 439 Convert binary $0000_2$ to $1111_2$ into two digit Binary Coded Decimal $00_{10}$ to $15_{10}$ (tens $f_4$, ones $f_3 f_2 f_1 f_0$).

## Transmission Gate TG

A transmission gate (TG) circuit uses pmos and nmos transistors as open (off) and closed (on) switches. A pmos transistor switch is on when the pmos gate voltage is L (0V, $V_{SS}$). And, an nmos transistor switch is on when the nmos gate voltage is H (1.8V, $V_{DD}$). *The on state is essentially independent of the source and drain voltages.*

An nmos transistor turns on when $V_{Gate\_to\_Source} > V_{TN}$ *or* $V_{Gate\_to\_Drain} > V_{TN}$ where $V_{TN}$ is the nmos threshold voltage. Analogous comments apply to the pmos transistor. A TG is a circuit with pmos only, or nmos only, or both. Here is a TG with both.

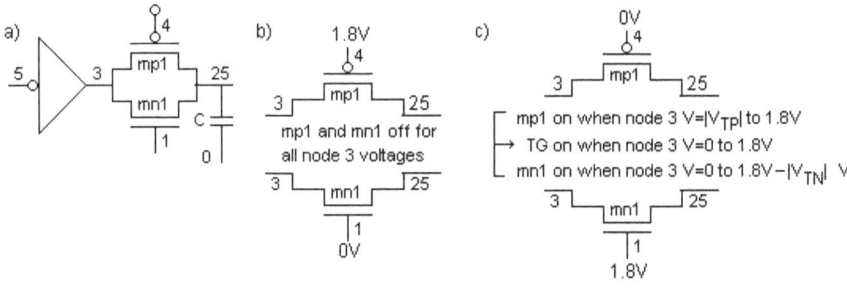

### 4.5.3 Manchester Full Adder[2]

A modern Manchester circuit uses transmission gates (TG sidebar p83) to implement the look ahead carry equations. The clocked version uses nmos TGs to propagate the active low carry (Figure 417a). The asynchronous version uses CMOS TGs to propagate the active high carry (Figure 417b). A true $g_n$ signal means carry out $c_n$ has to be generated. A true $p_n$ signal means carry $c_{n-1}$ into CMOS $TG_n$ has to be propagated to the node at the TG output to produce carry out $c_n$. A true $k_n$ signal means carry out $c_n$ is 0.

$(32a)$ $\quad c_n = g_n + p_n c_{n-1}$ $\qquad$ $(32b)$ $\quad k_n = x_n' y_n'$

**Figure 417 The Manchester Carry Chain Elements, two ways**

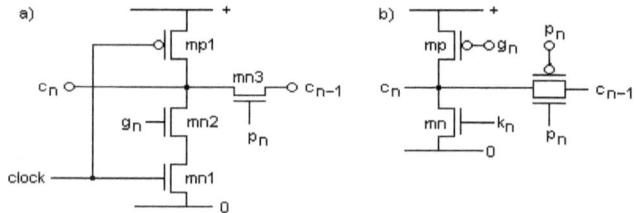

The $p_n$ signal turns on *CMOS $TG_n$* that connects node $c_{n-1}$ to node $c_n$. A true $g_n$ signal turns on $mp_n$ charging node $c_n$ to H. A true $k_n$ signal turns on $mn$

**Figure 418 The Manchester Carry Chain, 3 bits**

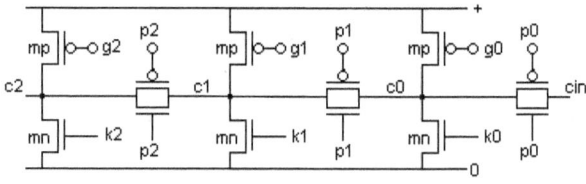

to discharge node $c_n$ to L so that carry out is a 0. The k's are referred to as kill signals. The chain implements the look ahead carry equations.

$(18a)$ $\quad c_0 = k_0'(g_0 + p_0 c_{in})$

$(18b)$ $\quad c_1 = k_1'(g_1 + p_1 c_0)$
$\qquad\quad = k_1'(g_1 + p_1 k_0'(g_0 + p_0 c_{in}))$
$\qquad\quad = k_1' g_1 + k_1' k_0' p_1 g_0 + k_1' k_0' p_1 p_0 c_{in})$

$(18c)$ $\quad c_2 = k_2'(g_2 + p_2 c_1)$
$\qquad\quad = k_2'(g_2 + p_2(k_1' g_1 + k_1' k_0' p_1 g_0 + k_1' k_0' p_1 p_0 c_{in}))$
$\qquad\quad = k_2' g_2 + k_2' k_1' p_2 g_1 + k_2' k_1' k_0' p_2 p_1 g_0 + k_2' k_1' k_0' p_2 p_1 p_0 c_{in}$

---

[2] IEE Vol 106 part B p464 Sept 1959.

In a Manchester Full Adder each bit pair produces g and p functions (Figure 419).

Each bit pair adds an identical circuit, and only one unit of delay to the carry chain.

**Figure 419 The Manchester Full Adder, 3 bits shown**

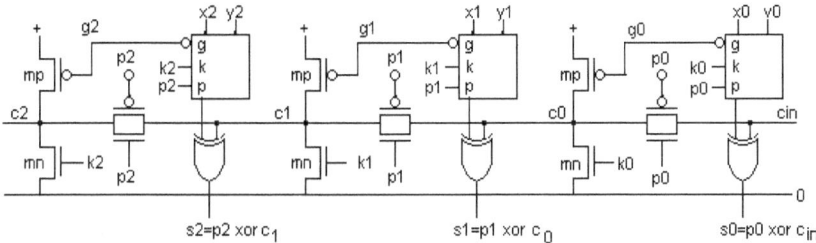

The adder is readily converted to a general purpose ALU (Figure 420) by addition of universal logic circuits such as the ULC (Section 4.5.2 page 82). The ALU is programmable via the ULC circuits that implement the equations for the g and p functions, and bit sums s. (The k, p, s codes are the same for all bit pairs.)

**Figure 420 A Manchester ALU, 1 bit**

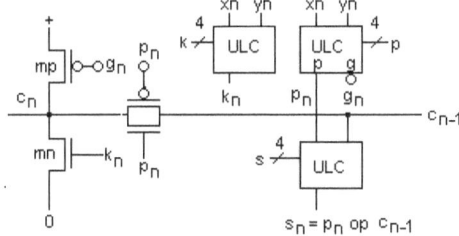

*Problem* 440 Reference Figure 420 and page 82. What are the k, p and s codes for all n?

*Problem* 441 Reference Figure 420. Explain why the circuit implements the carry equation $c_n=k_n'(g_n+p_nc_{n-1})$.

# 4.6 Comparing Numbers

When we compare positive base ten numbers 348 and 295 we can visually see that 348 is greater than 295. On the other hand digital circuits have to make logical tests. The symbols 348 and 295 are the unique names of the sums defining 348 (300+40+8) and 295 (200+90+5). The most significant digits (msd's) have the greatest value that is why comparisons start with the msd's. An msd test tells us that $3 > 2$. This is definitive, because $300 > 299 > 295$. A comparison process assumes both numbers have the same number of digits. If that is not so, then a comparator's most significant bit inputs are set as zeros as when comparing 54321 and 00039. Now compare 342 and 338. Digit by digit comparisons are 3=3, 4>3, 2<8. Comparisons move from msd to lsd. An algorithm comparing two positive numbers x and y starts with the most significant digit

Magnitude Comparison (*The standard comparator circuit is 85*).
1. Test a digit pair: the digits are equal or not equal.
2. IF the digits are equal
    IF this is the last digit pair assert the x=y output, and exit.
    ELSE move right to the next digit pair and go to step 1.
   ELSE (the digits are not equal)
    Assert the x>y or the x<y output lines, according to which is greater, and exit.

This algorithm requires a circuit to detect equality and another to detect which is greater than (Figure 421). Assemblies of these circuits are used to build comparators for x=y, x<y, and x>y (y>x and y<x output lines are not used). The less than AND circuit (x<y) is the same as the greater than circuit with x and y exchanged.

**Figure 421 Comparing two unsigned digits**

An xor output equals 0 when all inputs are equal to all zeros or all ones. In other words x=y, when the XOR output is 0. The truth table informs us that x<y when x'y=1.

| $y$ | $x$ | $yx'$ | $y \oplus x$ |
|---|---|---|---|
| 0 | 0 | 0 | 0 |
| 0 | 1 | 0 | 1 |
| 1 | 0 | 1 | 1 |
| 1 | 1 | 0 | 0 |

(19a)  $x_k$  equals  $y_k$  when  $0 = (x_k \oplus y_k)$

(19b)  $x_k$  is less than  $y_k$  when  $1 = x_k' y_k$   $(x_k = 0$  and  $y_k = 1)$

(20)  if  $x = x_3 x_2 x_1 x_0$  and  $y = y_3 y_2 y_1 y_0$

   then  $x = y$  when

   $$0 = (x_3 \oplus y_3) + (x_2 \oplus y_2) + (x_1 \oplus y_1) + (x_0 \oplus y_0)$$

(21)  *4 bit numbers*  $x < y$  when

   $(x_3 < y_3)$

*or*  $(x_3 = y_3)(x_2 < y_2)$

*or*  $(x_3 = y_3)(x_2 = y_2)(x_1 < y_1)$

*or*  $(x_3 = y_3)(x_2 = y_2)(x_1 = y_1)(x_0 < y_0)$

*Signed numbers* The most significant digits (msds) of *signed* numbers are the sign digits. The sign digit of a negative number is a 1. The sign digit of a positive number is a 0. Four possible comparisons of the most significant binary digit pair (x,y) and subsequent action are :

++ (0,0) magnitude compare x and y (report x=y, x>y, or x<y)

+− (0,1) report x>y

−+ (1,0) report x<y

−− (1,1) magnitude compare x and y (exchange x>y and x<y reports)

The rest of the comparison circuit is the same as the magnitude comparison circuit.

---

*Problem* 442 Show that an alternative equation for x=y where x = $x_1 x_0$ and y = $y_1 y_0$ is $(x_1 \oplus y_1)'(x_0 \oplus y_0)'$.

*Problem* 443 Show that the equation for *unsigned* x > y, where x = $x_1 x_0$, y = $y_1 y_0$ is $x_1 y_1' + (x_1 \oplus y_1)'(x_0 y_0')$

*Problem* 444 Design a circuit that compares two four bit unsigned numbers and makes > or = or < decisions.

---

# 4.7 Timing Hazards

Boolean equations represent *steady state* behavior. Boolean equations *imply* correct transient response to changes in inputs. In fact Boolean equations per se *do not guarantee* correct transient response, because Boolean equations do not account for propagation delays. A signal change at any circuit input requires a finite time to produce a new output, because the input signal change has to propagate through the circuit. Combinational logic circuit performance is measured by how quickly and reliably results are achieved. These are major reasons why time is an important issue in logic circuit design. Reliability and time are related, because unexpected outputs called *glitches* can occur during transients. Glitches arise from various combinations of propagation delays. Timing diagrams are used to analyze logic circuits for possible glitches. Before proceeding to timing diagrams consider propagation delay.

## 4.7.1 Propagation Delay

In the physical world nothing changes instantaneously, because light's velocity in free space is finite ($3\times10^8$ meters per second) and nothing travels faster than light. For us more useful forms for the velocity of light are 30cm/ns ($1ns=1\times10^{-9}$ seconds) and 300μm/ps ($1ps=1\times10^{-12}$ seconds). Furthermore, the velocity of light is reduced by $1/\sqrt{\varepsilon}$ in a medium whose (relative) dielectric constant $\varepsilon$ is greater than one. Magnetic mediums with (relative) magnetic permeability μ also reduce the velocity of light by $1/\sqrt{\mu}$. In an integrated circuit digital signals are electromagnetic waves that propagate at a reduced velocity of light (by $1/\sqrt{11.9}$) in the n or p type silicon medium. A signal switching from H to L, or vice versa, at one chip input terminal will cause a change at an output terminal after a finite amount of time elapses. Propagation delay gives rise to hazards and glitches in some circuits.

## 4.7.2 Hazards and Glitches

Hazards arise when an input change can reach an output *by more than one path* in the logic circuit. A hazard represents the possibility an unexpected output (a glitch) may occur. May, not will, occur because delays through circuits depend on circuit parameters that vary from one production lot to another.

> *Emphasis* Logic equations represent the steady state behavior of logic circuits. Changes in inputs induce transient behavior in logic circuits.

*Static hazards* Static hazards arise when two paths feed an AND circuit to produce an xx' term, or feed an OR circuit to produce a y+y' term. These terms arise when propagation delays make two complementary signals equal for short periods of time (e.g. in time interval 2 of Figures 422, 423).

For example, one implementation (Figure 422) shows how xx' produces a glitch and another (Figure 423) shows how y+y' produces a glitch. Assume the inverter has one unit of delay, and the AND, NAND circuits have two units of delay. Observe how the inverter delay makes two complementary signals equal during time interval 2 thereby producing a glitch in period 4.

**Figure 422 Glitch from xx'**

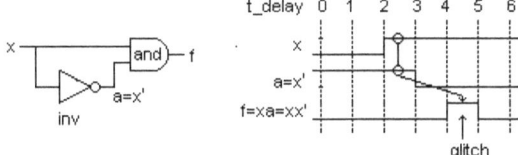

**Figure 423 Glitch from y+y'**

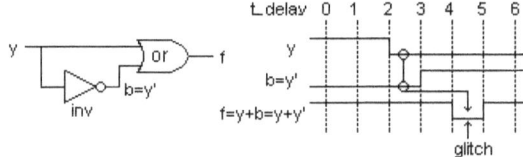

Now that we know how static hazards can manifest themselves as glitches we can examine equations for the presence of hazards.

$(22a)$ $\quad f = xy + x'z$

$(22b)$ $\quad if \ \ y = 1, \ \ z = 1, \ \ then \ \ f = x + x' \ \ (hazard)$

**Figure 424 f=xy+x'z**

If f is mapped onto a K map (Figure 424), then we see the two non overlapping clusters representing f. When x switches from 1 to 0 the xy cluster terms go to zero before the x'z cluster terms go to one. For a moment both clusters are zero, and the zero glitch results.

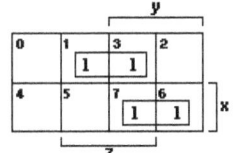

The consensus theorem indicates a solution. The consensus theorem (and the K map) tells us we can add the yz term and its corresponding cluster (Figure 425).

*the consensus theorem*

(23) $xy + x'z = xy + x'z + yz$

(24) $f = xy + x'z + yz$
    $if \ y = 1, \ z = 1 \ then \ f = x + x' + 1 = 1 \quad (no \ hazard)$

**Figure 425 f=xy+x'z+yz**

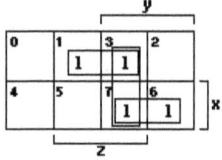

The yz cluster is independent of x (Figure 425), and so f cannot create a glitch when x switches from 1 to 0 or 0 to 1.

The key observation is that the glitch results because *the two clusters do not overlap*. Intuition tells us that an additional overlapping cluster will eliminate the hazard. This is true in general.

Clearly, overlapping clusters do not change the function because they do not add new 1 entries in the K map. The overlapping clusters are consensus terms.

***Dynamic hazards*** A dynamic hazard produces a glitch when an output changes more than once for each change in an input. For example, the ideal change in output from 0 to 1 is actually 0 1 0 1 or 1 0 1 0. Dynamic hazards are present when an input can reach an output via three or more paths that have different time delays. McClusky[2] explains how to discover dynamic hazards. Fortunately, it has been shown that eliminating static hazards, also eliminates dynamic hazards.

---

[2] McCluskey, E. J. Minimization of Boolean Functions. Bell System Technical Journal 35 (November 1956): 1417-1444.

# Review 4

Building blocks with no memory, implement logic equations as OR of ANDs or AND of ORs. When a building block has a memory it becomes a state machine (Chapter 5).

*Multiplexing* is a method by which N address lines select one out of $2^N$ input signals and send the selected signal to the multiplexer (mux) output. For example assume there are two input signals $a_1$, $a_0$. One address line z is required for selection, so that when z=0 input $a_0$ is selected and when z=1 input $a_1$ is selected. Important multiplexer applications are derived by making interpretations of the defining equations.

*Decoding* A product of variables used to select, i.e. assert, an output node in a logic circuit is referred to as a decoding function. Decoding is a method that makes output f=1 when the product of variables $a_K$ assigned to output f equals 1. Variables that are *inputs* to decoders are usually referred to as *address lines*. Decoders are minterm generators.

A decoder circuit is a logic circuit assembled from AND gates. A decoder has an enable input that activates the circuit. Address bits on address lines assert one output. Another view of a decoder is that the enable input is a signal input that is routed to the output selected by the address. Demultiplexer is another name for decoder.

*Binary Arithmetic Logic Circuits* Two is the radix of the binary number system. Radix 2 representations of numbers use the two digits 0 and 1 such as in the number $1101_2$ that is equivalent to decimal 13. Each digit of a binary number is readily represented in a two level logic system. The basic calculating functions are implemented by XOR and AND.

*Addition* The addition of large numbers is a straightforward extension of how one digit numbers are added. Add same-weight-digits and carry to the left. Same-weight-digits in two multi-bit numbers have the same weight: 1 or 2 or 4, and so forth. The addition table for one digit numbers is straightforward.

| 0 | 0 | 1 | 1 |
|---|---|---|---|
| +0 | +1 | +0 | +1 |
| 00 | 01 | 01 | 10 |

The sum function is XOR, and the carry function is AND. A half adder adds one column of bits and has a carry output. However, it does not have a carry input. A full adder is a half adder with provision for an input carry from the digit column at the right. The full adder equations are

$$sum = s = x \text{ xor } y \text{ xor } c_{in} \qquad carry = c_{out} = xy + c_{in}(x+y)$$

The ones column is implemented by a half adder, and full adders implement the remaining columns. The important question is When does the correct answer appear at the outputs? Assume the bits of both numbers are presented to the adder inputs simultaneously. Furthermore assume $t_d$ is the time to add two one digit numbers, and $t_d$ is also the time to produce a carry at the carry input of the column to the left. Then we can calculate the delays of the common scenarios ripple carry and look ahead carry. A ripple carry circuit adds one delay per bit. A more complex look ahead carry circuit eliminates accumulation of ripple carry delays by executing all bit carries at the same time for a one delay per sum.

*Universal Logic Circuit* The universal logic circuit (ULC) is an example of a circuit whose function is selected via programming bits. One ULC is designed to generate the 16 functions of two variables.

*Arithmetic Logic Unit (ALU 181)* Computer instructions perform logic and arithmetic operations. To minimize circuits Arithmetic Logic Units in computers perform both types of operations, because they can. For example, ALU units are used in computers to add (x+y) or subtract two numbers (x−y), add one to a number (x+1), form x AND y, x OR y, and pass x, or pass y unchanged through the ALU. Give or take some other functions this is a typical subset of the ULC codes.

*Comparing Numbers* When we compare decimal numbers 348 and 295 we can visually see that 348 is greater than 295. On the other hand digital circuits have to make digit by digit logical tests. XOR is used for equality, and AND for less than.

*Timing Hazards* Boolean equations represent steady state behavior. Boolean equations per se do not guarantee correct transient response, because they do not account for propagation delays. Reliability and time are related, because unexpected outputs called glitches can occur during transient behavior. Glitches arise from various combinations of propagation delays. Glitches are hazards that are eliminated by adding gates corresponding to consensus terms.

# 5 ASM State Machines

A state machine implements an algorithm. The ASM (algorithmic state machine) chart is a preferred way to represent the algorithm. A state machine has a next state combinatorial logic function, a state memory function, and an output combinatorial logic function (Figure 501).

*Next state combinatorial logic function* This function is implemented by building blocks with no memory and gates. External signals and state register outputs are inputs to this logic circuit. Outputs are inputs to the state memory function (state register flip-flops).

*State memory function* This function is implemented as an assembly of flip-flops. The set of flip-flops is referred to as a *state register* that stores a binary number representing the present state of the state machine. If the number is encoded, then the number of flip-flops is a minimum. If the number is not encoded, then there is one flip-flop per present state. For example, if there are six states three flip-flops encode the six numbers 0 to 5 (two states are not used). On the other hand six flip-flips are required to represent the numbers 0 to 5 when state numbers are not encoded.

**Figure 501 State Machine Block Diagram**

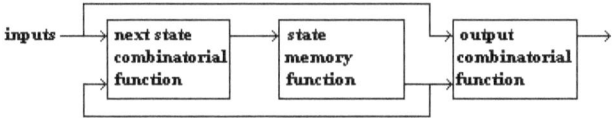

*Output combinatorial logic function* This function is implemented by building blocks with no memory and gates. External signals and state register outputs are inputs to this logic circuit. Outputs are the state machine's outputs.

*Unique next state* Given the *present state*, the *next state* must be determined without ambiguity for any values of state and input variables. Having arrived at the next state this next state becomes the present state.

> *The creative, and difficult, part of the design is producing the specification that describes the function(s) to be implemented. Implementation of the specification in the form of a synchronous state machine is straightforward.*

# 5.1 Algorithmic State Machine (ASM) Charts

The Algorithmic State Machine[1] [2] (ASM) chart represents a synchronous or an asynchronous finite state machine (FSM). The ASM chart (Figure 502 for example) is drawn using lines and three symbols: the *state rectangle*, the *input diamond*, and the *conditional output oval*. The finite state machine hardware design is derived directly from the Algorithmic State Machine chart (e.g. page 10). A synchronous machine's clock is not shown on the ASM chart in order to simplify the chart. States are represented by rectangles (S0, S1, S2, and S3 in Figure 502). The rectangle has only one entrance and one exit. Any number of paths may lead to a state rectangle's entrance. However, only one path leads away from the state rectangle's exit. The path may go directly to another state when this is an unconditional change of state. Or, the path enters a logic tree of conditions (one or more diamonds). Each condition has a corresponding logic circuit output (from the diamonds) to its next state.

**Figure 502 ASM Chart**

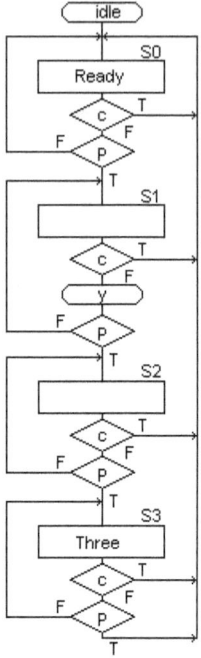

*State variable* State variables are the outputs of the state register flip-flops. For example, flip-flop outputs $q_1$ and $q_0$ are state variables.

*State* The state of a machine is the binary number representing the current combination of bits stored in the state register flip-flops such as $q_1 q_0$.

*State assignment* The state assignment process assigns a different number to each state to identify each state uniquely. The set of numbers is the state assignment. This is an important step, because the assignments affect the complexity of the next state circuit and output circuits. Know that assignments of numbers to states are not unique (5.5 page 102).

[1] Clare, C. R. 1971 Designing Logic Systems Using State Machines. ISBN 0701 111 200
[2] Winkel, D., and F. Prosser. 1987. The Art of Digital Design. ISBN 0130 467 804

*Synchronous Machine Clock* (not shown in the ASM chart). The clock action is as follows. At every rising clock edge the next state decision is made to step from the present state to the next state. The present state and the diamond's values (T or F) at the clock edges determine which state is the next state. For example, in Figure 502 when the present state is $S_2$, if c (clear) is True the next state is $S_0$. Follow the path from $S_2$'s exit through the c diamond's T-exit back to $S_0$'s entrance. When term c'p is True the T-exit path from the p diamond leads to $S_3$.

Notes: The flip-flop properties may dictate that every falling clock edge is used in lieu of every rising clock edge. The present state can be the next state (e.g. in Figure 502 $S_0$ steps to $S_0$ on every rising clock edge when c is True).

*Diamonds* represent branch decisions. External inputs are entered inside a diamond in the form of branch equations (e.g. c and p in Figure 502). The diamond symbol has True and False exit paths that implement a branch decision. The branch equation(s) placed in the diamond may be one or more Boolean functions of input variables. The extended diamond represents a tree of diamonds (Figure 703 page 126). A tree has one exit path for every output of the Boolean equations. The inputs to the branch equation may be inputs from the outside world, or they may originate from within the same state machine, or they may originate from another (linked) state machine (5.4 page 100). Note how the False output of the c diamond enters the p diamond (Figure 502, state $S_0$). The p diamond outputs represent $c'p'$ and $c'p$ decisions. This is a two diamond tree.

*Unconditional Branch* If no input variables (no diamonds) are associated with the present state, then the next state the machine enters is determined only by the present state.

*Conditional Branch* If one or more input variables (in diamonds) are associated with the present state, then the next state the machine goes to depends on this present state, the present values of the input variables, and the branch equation which is a function of the input variables. Input variables not associated with this present state have a don't care status when the state machine is in this present state.

*Outputs* Any state can be assigned no outputs, one or more unconditional outputs (in state rectangles), and, or one or more conditional outputs (in ovals). The same output can be unconditional in one state and conditional in another state.

***Unconditional outputs***[3] are listed in state rectangles. Unconditional outputs do not depend on inputs in any way. Unconditional outputs are asserted for every clock period the associated state is in the present state. For example, in Figure 502, *Three* is asserted for every clock period the state machine is in state $S_3$.

***Conditional outputs*** are shown in ovals that are always associated with a branch diamond representing the condition. Conditional outputs are asserted for every clock period the path from the present state, via one or more diamonds, to the oval entrance is asserted. For example, in Figure 502, the y conditional output is asserted *during* the present state when the present state is $S_1$, and c is False. In other words the conditional output depends on this present state, the present values of the input variables, and the branch equation which is a function of the input variables.

Note: Ovals, such as the one with the y output, are ignored when tracing paths to the next state.

***ASM (algorithmic state machine) chart*** The ASM chart is a network of states (rectangular boxes), each with or without branches (diamonds), unconditional outputs (labels in state rectangular boxes), and conditional outputs (labels in oval boxes). The ASM chart, such as the one in Figure 502, is our preferred way to represent any state machine algorithm.

***Timing diagrams*** A *timing diagram* depicts, in graphical format, circuit logic levels (H and L) as a function of time (Figure 508). The horizontal axis represents time, and the vertical axis represents logic levels (H or L). Time is measured in clock periods. A state is associated with each clock period. Each trace on a timing diagram shows when one logic variable changes level from high to low, and vice-versa.

**Figure 508 Timing diagram showing link actions**

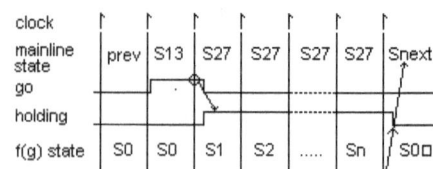

---

[3] The ASM includes traditional Moore machines which have only unconditional outputs and traditional Mealy machines which have only conditional outputs.

## 5.2 ASM to Truth Table

If you have a truth table, then you can design a circuit. If you have an ASM, then you can derive a truth table representing the ASM. Here is how you do that.

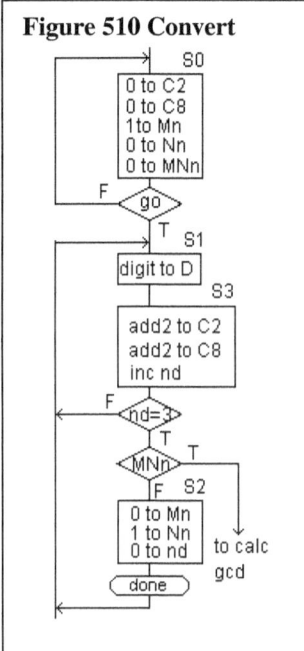

**Figure 510 Convert**

The truth table inputs are the state variables $q_1q_0$ (Figure 510), and the variables go, nd=3, MNn in the diamonds that make next state decisions. Table outputs are the next states $q_1^+q_0^+$, and unconditional and conditional ASM outputs as shown in the state boxes and ovals.

Truth table column headings are listed from left to right as present state, input variables, next state and ASM outputs.

| $q_1q_0$ | go | nd=3 | MNn | $q_1^+q_0^+$ | Outputs |
|---|---|---|---|---|---|

If a state x has one input variable associated with it then there are two possible next states. This is why two state x lines are entered into the table for state $00_2$. Two variables require 4 lines as in state $11_2$.

Here is the Convert truth table. Outputs are omitted.

Convert truth table

| $q_1q_0$ | go | nd=3 | MNn | $q_1^+q_0^+$ |
|---|---|---|---|---|
| 00 | 0 | - | - | 00 |
| 00 | 1 | - | - | 01 |
| 01 | - | - | - | 11 |
| 10 | - | - | - | 01 |
| 11 | - | 0 | 0 | 01 |
| 11 | - | 0 | 1 | 01 |
| 11 | - | 1 | 0 | 10 |
| 11 | - | 1 | 1 | Exit |

*Problem* 501, 502, 503 Make a truth table for each ASM.

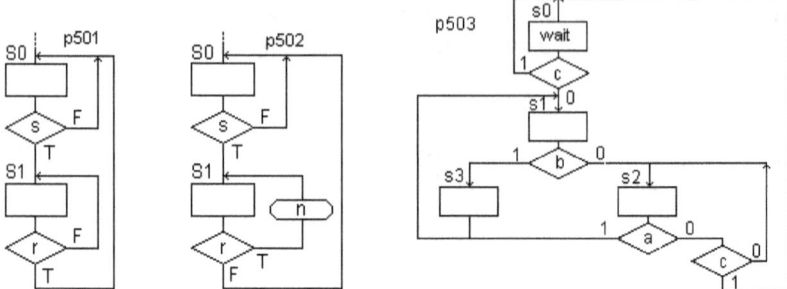

*Problem* 504 Make a truth table for ASM chart p504. Given the cba *waveforms* expressed as n-tuples, start in $S_2$ and plot the sequence of states.

| c | 11000000010 |
|---|---|
| b | 00000100000 |
| a | 11111100001 |
| *State* | 2 |

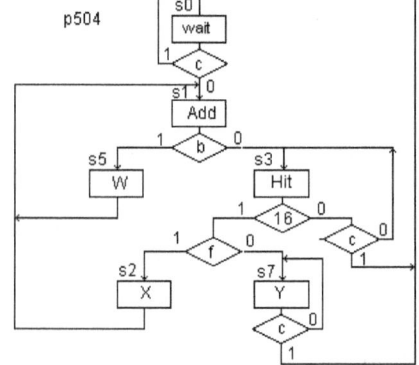

*Problem* 505 Draw the ASM chart that generates the sequence of states 01320132 etc. Use $q_1 q_0$ as state variables.

*Problem* 506 Make truth table for circuit p506. Draw the ASM chart for circuit p506 when in=1=H.

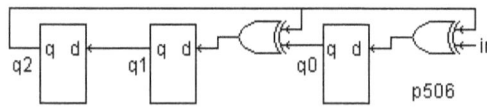

# 5.3 Synchronizing Inputs

An asynchronous input is not synchronized with the synchronous state machine clock. An asynchronous input transition that occurs *outside* the setup and hold window of a flip-flop may not be processed correctly. A related problem is the need to synchronize signals passing to and from two state machines that have different clock frequencies. Consider a state machine fragment (Figure 503).

**Figure 503 State machine fragment**

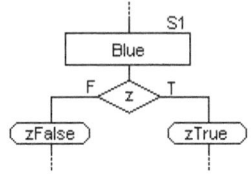

***Z is Asynchronous*** A timing diagram (Figure 504) shows how conditional outputs zFalse and zTrue can appear at the wrong time when input variable z is asynchronous.

***Z is Synchronous*** A timing diagram (Figure 505) shows how conditional outputs zFalse and zTrue appear correctly when input variable z is synchronous.

**Figure 504 Asynchronous input forces incorrect outputs**

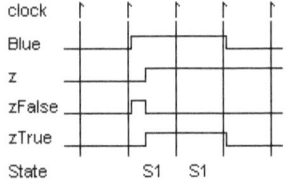

**Figure 505 Synchronous input**

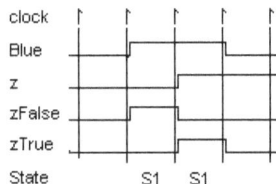

***Synchronizer circuit*** The first flip-flop in a suggested synchronizer circuit (Figure 506) can still go into a metastable state (Section 6.4 page 121).

**Figure 506 A synchronizer circuit**

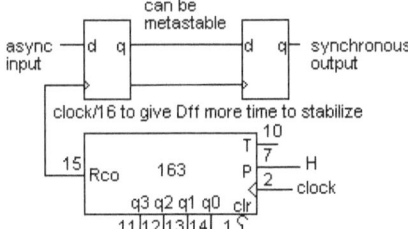

# 5.4 Linked State Machines

Linked state machines reduce the number of states required to implement a system. Consequently the hardware circuit has fewer parts. Any digital function that is used repeatedly by an algorithm may be implemented as a stand alone state machine that is linked to the main state machine. Linking allows repeated use of a function's circuit. This repeated use represents states not added to the mainline state machine. (Readers with computer programming experience will recognize that the linked state machine and its associated state machine are analogous to the subroutine.) Links are established when an output from state machine 1 is an input to state machine 2 and vice versa. A mainline state machine linked to state machines that are in turn linked to other state machines and so on is an example of top down design at its best. Consider an example (Figure 507).

**Figure 507 Mainline ASM linked to f(g) ASM**

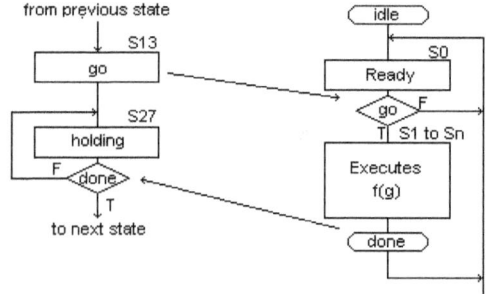

**Figure 508 Timing diagram showing link actions**

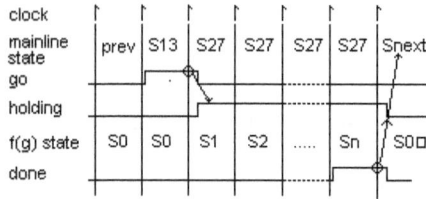

When the mainline ASM reaches state $S_{13}$ (Figures 507, 508) unconditional output *go* is asserted. On the next clock edge the mainline ASM moves to state $S_{27}$ that asserts the unconditional output *holding*. The mainline remains in $S_{27}$, because input done is not asserted. Output go is input go to the linked ASM that executes f(g). The f(g) ASM was holding in state $S_0$ waiting for input go to be asserted. Asserted input go allows the f(g) ASM to move to state $S_1$ that starts executing f(g). After some number of clock periods the f(g) ASM reaches state $S_n$ that asserts unconditional output *done* and returns to state $S_0$. Output done is input done to the mainline ASM. Asserted input done allows the mainline to exit from $S_{27}$ to the next state, whatever that happens to be.

---

*Problem* 507 Design the circuit specified by the linked ASM chart p507. Draw the timing diagram.

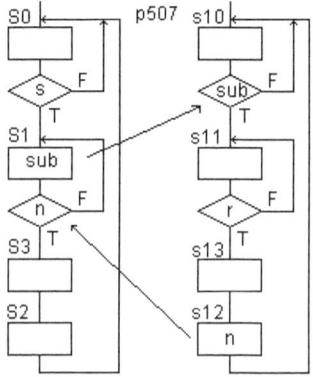

*Problem* 508 Design the circuit specified by the linked ASM chart p508. Draw the timing diagram.

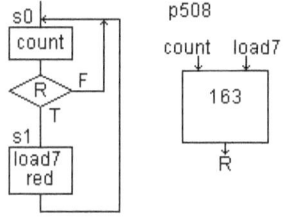

*Problem* 509 Design the circuit specified by the linked ASM chart p509. Draw the timing diagram.

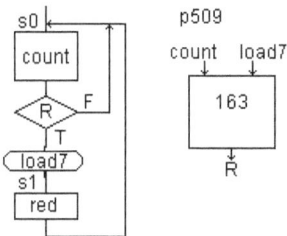

---

# 5.5 State Assignment and Races

State assignment affects the next state logic circuit and output circuit complexity. Furthermore state assignment may create transition races and output races that occur when the state machine changes state. Races are explained in upcoming paragraphs.

*State* The state of a machine is the binary number representing the current combination of bits stored in the state register flip-flops.

*Example*: The state of a state machine circuit with four flip-flops is represented by the set of the four flip-flop outputs $q_k$ concatenated in some order such as $q_3q_2q_1q_0$.

*State assignment* The state assignment process assigns a different number to each state to identify each state uniquely. The set of numbers is the state assignment. This is an important step, because assignments of numbers to states are not unique and state assignment impacts circuit complexity. The general problem of simultaneously considering all states has proven to be intractable to date. A practical method, *reduced dependency*, works on one state at a time. This method produces a reasonable solution.

*Reduced dependency* Clare[1] defined reduced dependency as *minimizing the number of state variables dependent upon an asynchronous input*. He showed one way to reduce dependency on asynchronous inputs such as x and y (Figure 509). If the present state $q_1q_0$ is 00 and $q_1^+q_0^+$ are dependent on variable x and x', the next states are 01 and 10. Whereas if only $q_1^+$ is dependent on y' then the next states are 01 and 11.

| | | $x$ | $q_1^+q_0^+$ | | | | $y$ | $q_1^+q_0^+$ |
|---|---|---|---|---|---|---|---|---|
| $q_1^+ = x'$ | $q_0^+ = x$ | 0 | 10 | | $q_1^+ = y'$ | $q_0^+ = 1$ | 0 | 11 |
| | | 1 | 01 | | | | 1 | 01 |

**Figure 509 Two ways to assign states**

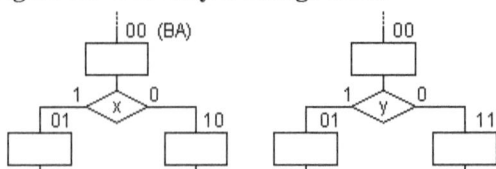

---

[1] Clare, C. R. 1971 Designing Logic Systems Using State Machines. ISBN 0701 111 200

***Transition Races*** A transition race can cause an output to occur at the wrong time, and perhaps transition the state machine to the wrong state. When two or more state variables change simultaneously a race can occur. The race is won by the variable that changes first. This motivates the following reduced dependency rule.

> *Only one state variable can depend on an asynchronous input in combination with any number of synchronous input variables*

***Output Races*** An output race can cause an output to occur at the wrong time or when it should not occur in the first place.

When two state variables change simultaneously $2^2$ or 4 possible state numbers may exist during the transition time initiated by the clock edge. If the state machine is in state 10 and the next state is supposed to be 01, then 00, 01, 10, and 11 state numbers may appear during the transition time. If an unconditional output is assigned to state 11 a false output glitch, or whatever, may occur during the transition time. The race is between the two state variables. Will the state number transition from 10 directly to 01 or will it transition from 10 to 11 to 01? *This emphasizes the importance of reduce dependency.*

***State Assignments simplified*** Here are several practical guidelines.
If one asynchronous input selects one out of two states as next states, then these two states must have adjacent state numbers. Numbers that are adjacent in the Karnaugh Map sense.

Gates are simplified when direct state transitions are assigned adjacent state numbers.

Gates are simplified when conditional state transitions have reduced dependency assignments.

Conditional outputs should not depend on asynchronous inputs, because then their duration is not synchronous.

Clock all output functions to avoid output races.

Check and recheck for races.

## 5.6 Euclid's gcd Calculator

The gcd calculation can be broken down into three parts – entering the digits of numbers M and N (page 147), converting the digits to binary numbers M and N (Figure 510), and calculating the gcd of M and N (Figure 511). The mathematics creates the ASM charts (Figures 510, 511).

*convert digits 6 7 8 9 into number 6,789*

$$6 \;\to\; 10\times6+7=67 \;\to\; 10\times67+8=678$$
$$\to\; 10\times678+9=6789$$

**Figure 510 Convert**

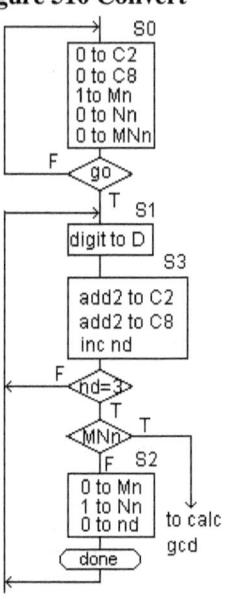

*Need to multiply by 10, which happens to equal*

$$8+2 = 2^3 + 2^1$$

*Shifting a number one bit to the left multiplies it by 2 with base 2 numbers so that* $10w = \left(2^3+2^1\right)w$

Here is $10w+x=(8+2)w+x$

|  |  |  |  |  |  |  |  |  |  |
|---|---|---|---|---|---|---|---|---|---|
| 0 | 0 | $w_3$ | $w_2$ | $w_1$ | $w_0$ | 0 |  | $2w$ | $C_2$ reg |
| $w_3$ | $w_2$ | $w_1$ | $w_0$ | 0 | 0 | 0 |  | $8w$ | $C_8$ reg |
| $a_6$ | $a_5$ | $a_4$ | $a_3$ | $a_2$ | $a_1$ | $a_0$ |  | $10w$ | $add_1$ out |
|  |  |  | $x_3$ | $x_2$ | $x_1$ | $x_0$ |  | $x$ | $D$ reg |
| $b_6$ | $b_5$ | $b_4$ | $b_3$ | $b_2$ | $b_1$ | $b_0$ |  | $10w+x$ | $add_2$ out |

Here is $10(10w+x)+y$

|  |  |  |  |  |  |  |  |  |  |  |
|---|---|---|---|---|---|---|---|---|---|---|
| 0 | 0 | $b_6$ | $b_5$ | $b_4$ | $b_3$ | $b_2$ | $b_1$ | $b_0$ | 0 | $2(10w+x)$ |
| $b_6$ | $b_5$ | $b_4$ | $b_3$ | $b_2$ | $b_1$ | $b_0$ | 0 | 0 | 0 | $8(10w+x)$ |
| $c_9$ | $c_8$ | $c_7$ | $c_6$ | $c_5$ | $c_4$ | $c_3$ | $c_2$ | $c_1$ | $c_0$ | $10(10w+x)$ |
|  |  |  |  |  |  | $y_3$ | $y_2$ | $y_1$ | $y_0$ | $y$ |
| $d_9$ | $d_8$ | $d_7$ | $d_6$ | $d_5$ | $d_4$ | $d_3$ | $d_2$ | $d_1$ | $d_0$ | $10(10w+x)+y$ |

Here is $10[10(10w+x)+y]+z=10[\alpha]+z$

|  |  |  |  |  |  |  |  |  |  |  |  |  |  |
|---|---|---|---|---|---|---|---|---|---|---|---|---|---|
| 0 | 0 | $d_9$ | $d_8$ | $d_7$ | $d_6$ | $d_5$ | $d_4$ | $d_3$ | $d_2$ | $d_1$ | $d_0$ | 0 | $2\alpha$ |
| $d_9$ | $d_8$ | $d_7$ | $d_6$ | $d_5$ | $d_4$ | $d_3$ | $d_2$ | $d_1$ | $d_0$ | 0 | 0 | 0 | $8\alpha$ |
| $e_C$ | $e_B$ | $e_A$ | $e_9$ | $e_8$ | $e_7$ | $e_6$ | $e_5$ | $e_4$ | $e_3$ | $e_2$ | $e_1$ | $e_0$ | $10\alpha$ |
|  |  |  |  |  |  |  |  |  | $z_3$ | $z_2$ | $z_1$ | $z_0$ | $z$ |
| $f_C$ | $f_B$ | $f_A$ | $f_9$ | $f_8$ | $f_7$ | $f_6$ | $f_5$ | $f_4$ | $f_3$ | $f_2$ | $f_1$ | $f_0$ | $10\alpha+z$ |

Example  w=9, x=9, y=9, z=9

| 1 | 2 | 3 | 4 | 5 | 6 | 7 | 8 | 9 | 10 | 11 | 12 | 13 | 14 | value | label |
|---|---|---|---|---|---|---|---|---|----|----|----|----|----|-------|-------|
| 0 | 0 | 0 | 0 | 0 | 0 | 0 | 0 | 0 | 0 | 0 | 0 | 0 | 0 | $2 \times 0$ | $C_2$ reg |
| 0 | 0 | 0 | 0 | 0 | 0 | 0 | 0 | 0 | 0 | 0 | 0 | 0 | 0 | $8 \times 0$ | $C_8$ reg |
| 0 | 0 | 0 | 0 | 0 | 0 | 0 | 0 | 0 | 0 | 0 | 0 | 0 | 0 | $10 \times 0$ | $add_1$ out o |
|   |   |   |   |   |   |   |   |   |   | 0 | 0 | 0 | 1 | 9 | $D$ reg |
| 0 | 0 | 0 | 0 | 0 | 0 | 0 | 0 | 0 | 0 | 1 | 0 | 0 | 1 | 9 | $add_2$ out |
| 0 | 0 | 0 | 0 | 0 | 0 | 0 | 0 | 0 | 1 | 0 | 0 | 1 | 0 | $2 \times 9$ | $C_2$ reg |
| 0 | 0 | 0 | 0 | 0 | 0 | 0 | 1 | 0 | 0 | 1 | 0 | 0 | 0 | $8 \times 9$ | $C_8$ reg |
| 0 | 0 | 0 | 0 | 0 | 0 | 0 | 1 | 0 | 1 | 1 | 0 | 1 | 0 | $10 \times 9$ | $add_1$ out |
|   |   |   |   |   |   |   |   |   |   | 1 | 0 | 0 | 1 | 9 | $D$ reg |
| 0 | 0 | 0 | 0 | 0 | 0 | 0 | 1 | 1 | 0 | 0 | 0 | 1 | 1 | 99 | $add_2$ out |
| 0 | 0 | 0 | 0 | 0 | 0 | 1 | 1 | 0 | 0 | 0 | 1 | 1 | 0 | $2 \times 99$ | $C_2$ reg |
| 0 | 0 | 0 | 0 | 1 | 1 | 0 | 0 | 0 | 1 | 1 | 0 | 0 | 0 | $8 \times 99$ | $C_8$ reg |
| 0 | 0 | 0 | 0 | 1 | 1 | 1 | 1 | 0 | 1 | 1 | 1 | 1 | 0 | $10 \times 99$ | $add_1$ out |
|   |   |   |   |   |   |   |   |   |   | 1 | 0 | 0 | 1 | 9 | $D$ reg |
| 0 | 0 | 0 | 0 | 1 | 1 | 1 | 1 | 1 | 0 | 0 | 1 | 1 | 1 | 999 | $add_2$ out |
| 0 | 0 | 0 | 1 | 1 | 1 | 1 | 1 | 0 | 0 | 1 | 1 | 1 | 0 | $2 \times 999$ | $C_2$ reg |
| 0 | 1 | 1 | 1 | 1 | 1 | 0 | 0 | 1 | 1 | 1 | 0 | 0 | 0 | $8 \times 999$ | $C_8$ reg |
| 1 | 0 | 0 | 1 | 1 | 1 | 0 | 0 | 0 | 0 | 0 | 1 | 1 | 0 | $10 \times 999$ | $add_1$out |
|   |   |   |   |   |   |   |   |   |   | 1 | 0 | 0 | 1 | 9 | $D$reg |
| 1 | 0 | 0 | 1 | 1 | 1 | 0 | 0 | 0 | 0 | 1 | 1 | 1 | 1 | 9999 | $add_2$out |

*Calculate the gcd* – Division is done by the straightforward use of repeated subtraction until R<N. Then Move N to M and R(=M−N) to N.

$M - N = M + N' + 1$ *(the negative of a number is the complement plus 1)*
$M = 585_{10} = 1001001001_2 \quad N = 165_{10} = 0010100101_2 \rightarrow N' + 1 = 1101011011_2$

| | | | | | | | | | | | | |
|---|---|---|---|---|---|---|---|---|---|---|---|---|
| 1 | 0 | 0 | 1 | 0 | 0 | 1 | 0 | 0 | 1 | 585 | $M$ | 585 |
| 1 | 1 | 0 | 1 | 0 | 1 | 1 | 0 | 1 | 1 | −165 | $N' + 1$ | +835 |
| 0 | 1 | 1 | 0 | 1 | 0 | 0 | 1 | 0 | 0 | 420 | $M - N$ | 420 |

**Figure 511 Partial ASM convert and calculate gcd**

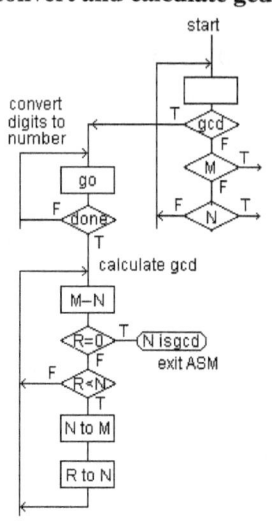

| | | | | | | | | | | | | |
|---|---|---|---|---|---|---|---|---|---|---|---|---|
| 0 | 1 | 1 | 0 | 1 | 0 | 0 | 1 | 0 | 0 | 420 | $M$ | 420 |
| 1 | 1 | 0 | 1 | 0 | 1 | 1 | 0 | 1 | 1 | −165 | $N' + 1$ | +835 |
| 0 | 0 | 1 | 1 | 1 | 1 | 1 | 1 | 1 | 1 | 255 | $M - N$ | 255 |

| | | | | | | | | | | | | |
|---|---|---|---|---|---|---|---|---|---|---|---|---|
| 0 | 0 | 1 | 1 | 1 | 1 | 1 | 1 | 1 | 1 | 255 | $M$ | 255 |
| 1 | 1 | 0 | 1 | 0 | 1 | 1 | 0 | 1 | 1 | −165 | $N' + 1$ | +835 |
| 0 | 0 | 0 | 1 | 0 | 1 | 1 | 0 | 1 | 0 | 90 | $M - N$ | 090 |

*Since $M - N = R = 90 < N = 165$*
*we move N to M, and R to N*

| | | | | | | | | | | | |
|---|---|---|---|---|---|---|---|---|---|---|---|
| 0 | 0 | 1 | 0 | 1 | 0 | 0 | 1 | 0 | 1 | 165 |
| 1 | 1 | 1 | 0 | 1 | 0 | 0 | 1 | 1 | 0 | −90 |
| 0 | 0 | 0 | 1 | 0 | 0 | 1 | 0 | 1 | 1 | 75 |

| | | | | | | | | | | | |
|---|---|---|---|---|---|---|---|---|---|---|---|
| 0 | 0 | 0 | 1 | 0 | 1 | 1 | 0 | 1 | 0 | 90 |
| 1 | 1 | 1 | 0 | 1 | 1 | 0 | 1 | 0 | 1 | −75 |
| 0 | 0 | 0 | 0 | 0 | 0 | 1 | 1 | 1 | 1 | 15 |

*continue by repeated subtraction of 15*
$75 - 15 = 60$
$60 - 15 = 45$
$45 - 15 = 30$
$30 - 15 = 15$
$15 - 15 = 0 \qquad$ *gcd is 15*

The mathematical process converting digits to a number specifies the convert part of the ASM, which we set up as an ASM linked to the main ASM. The process shows the need for $C_2$ and $C_8$ registers for times 2 and times 8 of a number by proper placement of the bits to be multiplied by 2 and 8. The $Add_1$ adder adds numbers $C_2$ and $C_8$. The next digit is stored in the D register. The $Add_2$ adder adds digit in D to the $C_2$, $C_8$ sum in $Add_1$ (Figure 722).

**Figure 510 Convert**

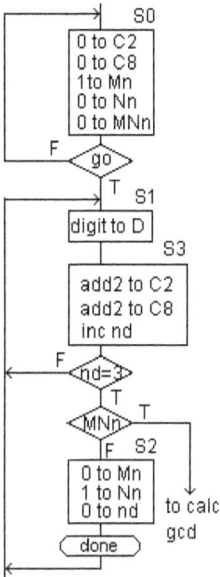

In state $S_0$ $C_2$ and $C_8$ are initialized to zero, and variables are initialized as *Mn=1, Nn=0, MNn=0.* Earlier, in the main ASM, digit count *nd* was set to 0. Mn=1 means the digits of M will be converted to a number. A *go* from the main ASM steps the linked convert ASM to $S_1$ where the first M digit *w* is placed in register D. The $Add_2$ sum is now 0+0+w, because this is a no memory circuit that produces the sum from the present values. The ASM moves to $S_3$ where $Add_2$ is loaded into $C_2$ and $C_8$, and *nd* is incremented to 1. Since *nd=3* is false the ASM moves to $S_1$ where *x* is stored in D and the $Add_2$ sum is now 10(10w+x). When *nd=3* is true M has been converted and in $S_2$ N is set up to be converted.

*Calculate the gcd* The mathematical process calculating the gcd defined the gcd calculating part of the ASM (Figure 511). The ASM calculating the gcd implements the gcd algorithm. Division of M by N is done by the straight-forward use of repeated subtraction until $0 \leq R < N$. Then move N to M and R to N.

**Figure 722 Convert Datapath**

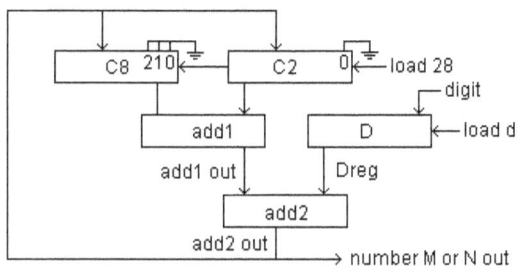

Divide the new M by the new N until the new R is less than the new N. Proceed until R=0.

# 6 One Bit Building Blocks with Memory

Building blocks with memory are known as *sequential circuits*. We prefer the modern name *state machine*. Present *output values* depend on the *past values stored in the memory*, and the *present input values*. When a present input value changes state only some output values change after a delay. When a clock changes state, new stored values are produced after a delay. (A clock is just another input.) Then new values appear at the outputs after another delay computing them from present inputs and the new stored values. State machines with clock inputs are referred to as *synchronous state machines. Asynchronous state machines*, which we do not discuss, do not use clocks.

Memory is implemented by *feedback* of a circuit output to a circuit input. The SR Latch is the most elementary one-bit state machine with memory. Latches do not have clock inputs. Edge Triggered flip-flops are one-bit memory circuits with clock inputs. They are *synchronous* circuits. *A flip-flop's next state defining equation is the most important fact a designer needs to know*. The ideas of *Setup and Hold Windows* guarantee correct *synchronous* circuit operation. We discuss why Metastability can create disasters in digital circuits.

Many complex functions are designed to be implemented as a sequence of events. By sequence we mean time elapses as events occur. Furthermore, the sequence may vary according to imposed conditions such as if the bell rings then the red light will turn on, else the yellow light will turn on. The circuit remembers that the bell did or did not ring, and proceeds accordingly. Circuits that implement sequences of events must remember events that occurred in the past.

A modern synchronous state machine specification process is presented in Chapter 5 where we show how to specify the state machine in the form of an ASM chart and how to derive the (design) truth table from the chart. Given the truth table the circuit design is straightforward. An example is on pages 7, 10, 11. Additional examples are in Chapter 7.

In order to design and build complex state machines we need to know about *one bit building blocks with memory*. We start with the SR latch.

# 6.1 SR Latch, an Asynchronous Circuit

A logic circuit acquires the memory property when a circuit output is connected to a circuit input. This connection from an output to an input is known as *feedback*. The SR Latch (set-reset latch) is the most elementary logic circuit with the memory property. The SR latch is a two gate circuit. Starting with a NAND gate, *pulse s* at the input is inverted and replicated at the *output q* after a time delay $t_D$ referred to as a *propagation delay* (Figure 601a). Propagation delay is the time required for changes at an input to appear at an output. (*The standard sr latch circuit is 75.*)

The *timing diagram* (Figure 601a) shows that the output signal q changes from L to H some $t_D$ nanoseconds after the set input s changes from H to L. In general the time delays for L to H and H to L transitions are different. However, the difference is a second order effect we will not concern ourselves with.

**Figure 601 SR Latch using NAND gates**

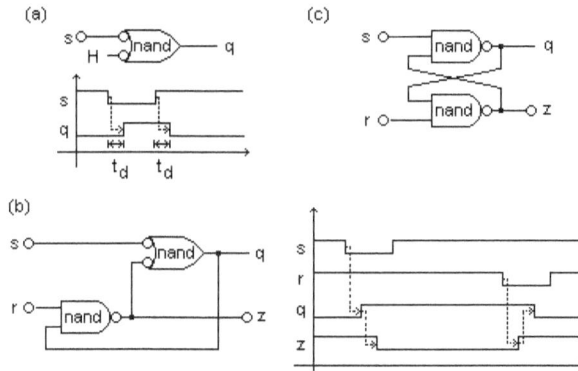

When output q follows input s (Figure 601a), we say the circuit is *transparent*. However when feedback is introduced, by connecting an output signal to an input (Figure 601b), the circuit is no longer transparent. Output q does not follow s if the latch is in the set state. The associated timing diagram shows circuit action when a pulse appears at the s (set) input, and later, shows circuit action when a pulse appears at the r (reset) input. The diagram shows propagation delays for each gate.

Assume s and r are not asserted (they are both H). Also assume q is L so that the latch is in the reset state.

Digital Design

***Set*** When s asserts and goes L, q is driven H (Figure 601b). Then with r at H and q at H, z is driven L to hold q at H. Later s returns to H and q remains at H, because z is L. The latch is now set.

***Reset*** Later, when r asserts and goes L, z is driven H (Figure 601b). Then with both s and z at H, q is driven L to hold z at H. Later r returns to H and q remains at L, because z is still H. The latch is now reset.

Note: s and r signals can be steps. They do not have to be pulses.

This SR latch assembled from NANDs is usually drawn as shown in Figure 601c. The mixed logic schematic (Figure 601b) is more easily understood. When opposite assertion levels are required, use an SR latch with NOR gates (Figure 602).

**Figure 602 SR Latch using NOR gates**

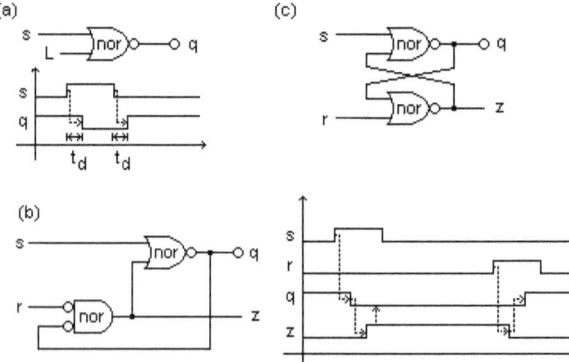

The world says the SR latch has two stable states (it is *bistable*). The stable states have the names set and reset. The *defining equation* is derived by circuit analysis. Cut the feedback line.

Then $q_{out} = s + z$ and $z = r'q_{fdbk}$

substituting for $z$: $q_{out} = s + r'q_{fdbk}$

$q_{fdbk}$ is a circuit input. Call it $q$.

$q_{out}$ appears after propagation delay $t_p$. Call it $q^+$.

Reconnect the feed back line. The SR latch defining equation is

$$\boxed{q^+ = s + r'q}$$

110

Suppose one or more of the input variables r, s change. What is q? One answer uses the right now (present) values of r, s, and q to evaluate s+r'q. After changes to inputs propagate through the circuit the output q assumes the calculated value. This is why *the new value of q is called $q^+$ (q plus)*. The q on the left side of the equation does not equal the q on the right side. To repeat: $q^+$ *is the next state of the present state q* that results after signal propagation times expire after s or r inputs change.

The r and s inputs are said to provide *excitation* for the latch. (In the past the truth table representing the defining equation was referred to as an excitation table.) The truth table is derived from the SR latch defining equation. The next state K map for the SR latch is shown in Figure 603.

**Figure 603 Next state K map for the SR latch**

| r | s | $q^+$ |
|---|---|---|
| 0 | 0 | q (*hold*) |
| 0 | 1 | 1 (*set*) |
| 1 | 0 | 0 (*reset*) |
| 1 | 1 | 1 (*do not use*) |

If *both r and s are asserted and change simultaneously*, the next state might be set, reset, or the latch could oscillate (no stable state). This is what *do not use* means. In practice, the final value is determined by the last input to remain asserted.

The SR latch is an asynchronous circuit with no clock input.

> *One condition for correct asynchronous circuit operation is that only one input changes at any time.*

---

*Problem* 601 Use the NAND SR latch in Figure 601. Plot q for the $s_L$, $r_L$ inputs. Mark each interval between events as set S, reset R, or hold H.

$s_L$ ⊓_⊔___⊓_⊔_____⊓_⊔_____

$r_L$ ____⊔_____⊔___

---

# 6.2 Synchronous Circuits

A logic circuit with no clock inputs is asynchronous. The outputs respond immediately to input changes. By immediately we mean after propagation delays elapse. An asynchronous circuit may or may not include feedback from one or more outputs to one or more inputs. I.e. an asynchronous circuit may or may not have memory. The SR latch (6.1) is an example of an asynchronous circuit with memory. Any combinatorial circuit is an example of an asynchronous circuit with no feedback. We will learn that

> *Reliable asynchronous and synchronous circuits require (1) that only one input at a time change, and (2) that the circuit be in a stable condition before any input changes.*

Stable condition means there are no ongoing transients anywhere in the circuit. All nodes are either at H or at L voltage. In the practical world these conditions may be difficult to achieve. This reality raises the question: *How do we satisfy conditions (1) and (2)?*

## 6.2.1 Clock

A clock is an independent signal source that emits a periodic signal. In a digital system the clock waveform is essentially a squarewave (Figure 604a). The transition from L to H is referred to as the positive edge (marked as up arrows in Figure 604c), and the transition from H to L is referred to as the negative

**Figure 604 Periodic Clock Waveform**

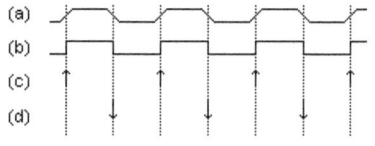

edge (marked as down arrows in Figure 604d). The transition times are referred to as the rise and fall times respectively. Idealized waveforms with zero transition times make timing diagrams more readily perceptible (Figure 604b).

The time between two consecutive positive, or two consecutive negative, transitions is the clock's *period T* that is the reciprocal of the clock's *frequency f*. A 500MHz clock has a 2ns period (T=1/f). We emphasize that there is one *triggering edge* per period by showing the clock waveform as a series of up arrows (Figure 604c), or down arrows (Figure 604d).

*Aperiodic edges* For example, the periodic clock can be shut off. Then a special circuit allows one pulse, with one triggering edge, to be emitted by one press of a push button (Experiment 9 page 279).

### 6.2.2 The Setup and Hold Window

A logic circuit with a clock whose positive or negative edges implement changes in state is a synchronous circuit. The selected clock edges (up or down but not both) are the *active clock edges or triggering edges.*

Asynchronous circuits have constraints:
1. Only one input can change at a time
2. The circuit must be in a stable condition during an input transition.

An asynchronous circuit with at least one input designated as a clock is a synchronous circuit. This is why synchronous circuits have the same constraints as asynchronous circuits. These constraints are satisfied if there are no input changes inside *the setup and hold window* (Figure 605). Only the clock changes inside the window.

**Figure 605 Setup and Hold Window**

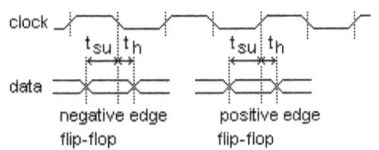

The setup time ($t_{SU}$) is a quiet time *before* an active clock edge. The setup time is the time interval before the clock transition during which no inputs should change, and any internal transients have died out.

The hold time ($t_H$) is a quiet time *after* an active clock edge. The hold time is the time interval after the clock transition during which no inputs should change, and any internal transients have died out.

The actual numerical value for setup time is determined by how long a circuit requires to dissipate transients created by any input change before the active clock edge transitions. The active clock edge also creates transients in the circuit. The hold time allows time for these transients to dissipate, and to achieve the correct result before new input changes start up new transients. Thus setup and hold times guarantee that if only the clock changes inside, and only other inputs change outside, the setup and hold window, then the circuit will function correctly. Setup time and hold time values that are technology dependent are found in data books.

# 6.3 The Edge Triggered Flip-Flop

An asynchronous logic circuit is converted to a synchronous logic circuit when you add bistable memory elements whose outputs are changed only by active clock edges. A synchronous bistable memory element is called a *flip-flop* (an asynchronous bistable memory element is called a *latch*). If an active clock edge flips an output from L to H, then some other active clock edge can flop it back to L. Other inputs to the flip-flop determine which clock edges flip and flop a flip-flop. Specifically we present *edge triggered flip-flops*.

At the end of a long design and development cycle a variety of flip-flop designs narrowed down to two major commercially available flip-flops: the D flip-flop and the JK flip-flop. A flip-flop has two states known as set and reset. The output's symbol is q. The output q reports the state of the flip-flop. *In a state machine a flip-flop's q is called a state variable.*

> *A flip-flop's next state defining equation is the most important fact a designer needs to know.*

Everything is derived from the defining equation. The t flip-flop is a useful special case of the jk flip-flop where j=k=t.

---

*Defining equations* :

d flip-flop: $q^+ = d$

jk flip-flop: $q^+ = jq' + k'q$

t flip-flop: $q^+ = tq' + t'q = t \ xor \ q$

---

### 6.3.1 D positive edge triggered flip-flop

The D flip-flop ($D_{FF}$) building block has a synchronous input clock, and data input d. The d input defines the next state (Figure 606). Asserting the asynchronous preset or clear inputs sets or resets the q outputs, while overriding the synchronous inputs. While the asynchronous preset or clear inputs are not asserted, the defining equation is operational. The (present state) data at the d input meeting setup and hold requirements are transferred to the (next state) output q on the next positive clock edge (Figure 607 page 116). Triggering occurs when the edge rises up past a voltage level, and is not dependent on edge's rise time over a wide range. After the hold time interval elapses data d may be changed with no effect on the outputs.

The $D_{FF}$ has two output lines for q. One output is active high and the other is active low. The truth table for the $D_{FF}$ includes inputs pre, clr, d, and the clock. The up arrow symbol means the defining equation is executed on the positive edge of the clock. (*The standard d flip-flop circuit is 74.*)

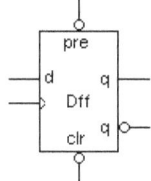

**Figure 606 Positive edge triggered D flip-flop**

| $D_{ff}$ Inputs | | | | Outputs | |
|---|---|---|---|---|---|
| *pre* | *clr* | *clk* | *d* | *q* | *q'* |
| L | L | – | – | * | * |
| L | H | – | – | H | L |
| H | L | – | – | L | H |
| H | H | ↑ | L | L | H |
| H | H | ↑ | H | H | L |
| H | H | L | – | $q_o$ | $q'_o$ |

\* *both H but unstable,*

$q_o$ *is present state, q is next state* $q^+$

The first line of the $D_{FF}$ truth table reveals the indefinite SR latch behavior when both pre and clr are active low. The dashes mean don't care.

| d | $f_d$ |
|---|---|
| 0 | 0 |
| 1 | 1 |

This is the d defining equation. $\boxed{f_d = d}$

*D flip-flop timing diagram* If any clock period n is the current period then it is the *present period*, and clock period n+1 is the next period. The value of d in the present period is the *present state of d*. Since $q^+=d$ the output in clock period n+1 is the *next state of q* that is a copy of present state d (Figure 607). Thus the q waveform is the same as the d waveform delayed in time by one clock period. (A one clock period slip assumes the d input is synchronous with the clock.) We do not show a $q^+$ waveform, because in the next state $q^+$ becomes the present state q.

**Figure 607 D flip-flop timing diagram**

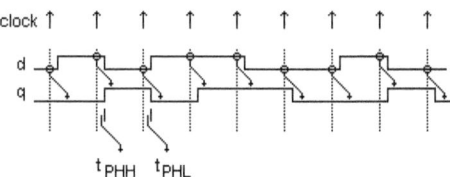

The coincidence of the d values and the clock edge are marked by circles (Figure 607). The circles mark the cause and arrows show the effect.

The propagation delays from clock to output must be greater than zero. When $t_{PHL}$ and $t_{PLH}$ (Figure 607) are not greater than zero, then the d and q waveform transitions merge with the clock edge. Then equation $q^+=d$ cannot answer the question *What is the value of d at the clock edge?* When propagation delays are zero there is no definite answer. This is why propagation delays greater than zero are necessary for synchronous operation. Since all physical circuits have propagation delays greater than zero this is not a problem.

---

*Problem* 602 Plot $q_0$, $q_1$ for the x, y inputs.

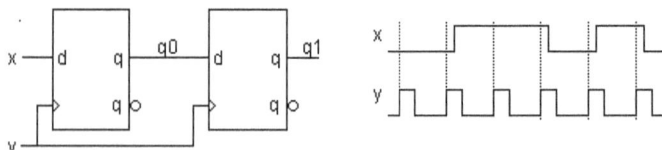

---

*Problem* 603 Starting with all flip-flop outputs at the L level plot $d_0$, $q_0$, $q_1$ over seven clock periods.

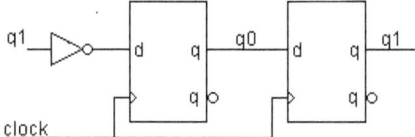

*Problem* 604 Starting with all flip-flop outputs at the H level plot $d_0$, $q_0$, $d_1$, $q_1$ over six clock periods.

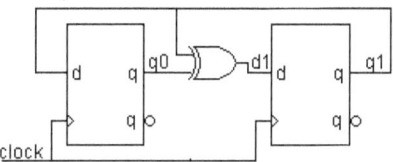

*Problem* 605 Starting with all flip-flop outputs at the H level plot $q_0$, $q_1$, $d_2$, $q_2$ over nine clock periods.

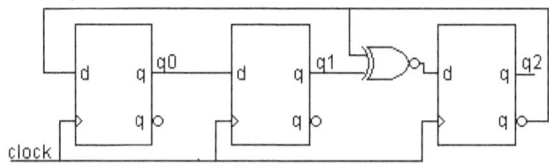

*Problem* 606 Derive the equations for flip-flop parameters $d_0$, $d_1$, $q_0^+$, $q_1^+$ of the circuit in problem 602. List all present states defined as $q_1q_0$, and all corresponding next states $q_1^+q_0^+$.

*Problem* 607 Derive the equations for flip-flop parameters $d_0$, $d_1$, $q_0^+$, $q_1^+$ of the circuit in problem 603. List all present states defined as $q_1q_0$, and calculate the corresponding next states $q_1^+q_0^+$.

*Problem* 608 Derive the equations for flip-flop parameters $d_0$, $d_1$, $q_0^+$, $q_1^+$ of the circuit in problem 604. List all present states defined as $q_1q_0$, and calculate the corresponding next states $q_1^+q_0^+$.

*Problem* 609 Derive the equations for flip-flop parameters $d_0$, $d_1$, $d_2$, $q_0^+$, $q_1^+$, $q_2^+$ of the circuit in problem 605. List all present states defined as $q_2q_1q_0$, and calculate the corresponding next states $q_2^+q_1^+q_0^+$.

## 6.3.2 JK positive edge triggered flip-flop

The JK flip-flop ($JK_{FF}$) has synchronous input clock, and data inputs j, k. The j and k inputs define the next state (Figure 608). Asserting the asynchronous preset or clear inputs sets or resets the q output, while overriding the synchronous inputs. While the asynchronous preset or clear inputs are not asserted, the defining equation is operational. The (present state) data at the j, k inputs meeting setup and hold requirements are transferred to the (next state) output q on the next positive clock edge. Triggering occurs when the edge rises up past a voltage level, and is not dependent on edge's rise time over a wide range. After the hold time interval elapses inputs may be changed with no effect on the outputs. (*The standard jk flip-flop circuit is 109*)

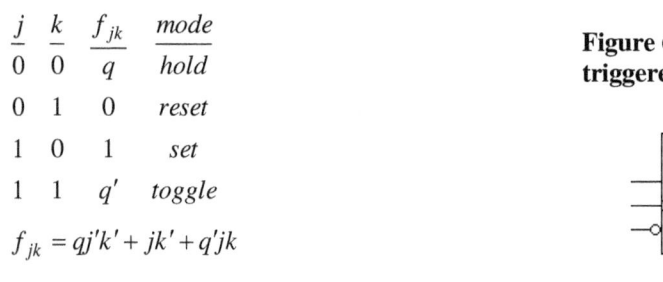

| j | k | $f_{jk}$ | mode |
|---|---|---|---|
| 0 | 0 | q | hold |
| 0 | 1 | 0 | reset |
| 1 | 0 | 1 | set |
| 1 | 1 | q' | toggle |

$$f_{jk} = qj'k' + jk' + q'jk$$

**Figure 608 Positive edge triggered JK flip-flop**

This is the jk defining equation. $\boxed{f_{jk} = jq' + k'q}$

If the present state q is zero then $f_{jk}=j$ so that k is a don't care input. The next state depends only on j. If the present state q is one then $f_{jk}=k'$ so that j is a don't care input. The next state only depends on k'. Then the jk excitation table above can be recast as a very useful state transition table.

| State | j | k | Transition |
|---|---|---|---|
| 0 | 0 | - | 0 to 0 |
| 0 | 1 | - | 0 to 1 |
| 1 | - | 1 | 1 to 0 |
| 1 | - | 0 | 1 to 1 |

*Problem* 610 Use the JK ff. Plot q. Each sequence of states starts in the reset state (q=0).

a)
$$J \quad 01000100000$$
$$K \quad 00010000100$$

b)
$$J \quad 0100000000$$
$$K \quad 1111100100$$

c)
$$J \quad 11000111111$$
$$K \quad 00010000100$$

*Problem* 611 Starting with all flip-flop outputs at the L level plot $q_0$, $q_1$, over six clock periods.

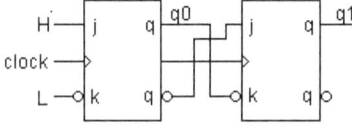

*Problem* 612 Starting with all flip-flop outputs at the L level plot $q_0$, $j_1$, $q_1$, over six clock periods.

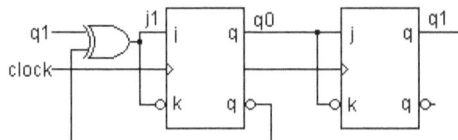

*Problem* 613 Starting with all flip-flop outputs at the L level plot $q_0$, $q_1$, $q_2$ over eight clock periods.

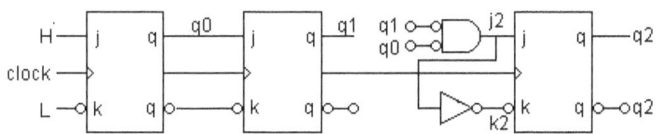

*Problem* 614 Derive the equations for flip-flop parameters $j_0$, $k_0$, $j_1$, $k_1$, $q_0^+$, $q_1^+$, of the circuit in problem 611. List all present states defined as $q_1 q_0$, and calculate the corresponding next states $q_1^+ q_0^+$.

*Problem* 615 Derive the equations for flip-flop parameters $j_0$, $k_0$, $j_1$, $k_1$, $q_0^+$, $q_1^+$, of the circuit in problem 612. List all present states defined as $q_1 q_0$, and calculate the corresponding next states $q_1^+ q_0^+$.

*Problem* 616 Derive the equations for flip-flop parameters $j_0$, $k_0$, $j_1$, $k_1$, $j_2$, $k_2$, $q_0^+$, $q_1^+$, $q_2^+$ of the circuit in problem 613. List all present states defined as $q_2 q_1 q_0$, and calculate the corresponding next states $q_2^+ q_1^+ q_0^+$.

Digital Design

## 6.3.3 T positive edge triggered flip-flop

The T flip-flop is a special case of the JK flip-flop where t=j=k.

The t defining equation is $q^+ = tq' + t'q = t\ xor\ q$ where $t = j = k$

| row | t | q | $q^+$ | mode |
|-----|---|---|-------|------|
| 0 | 0 | 0 | 0 | $q^+ = q$ hold $q$ when $t = 0$ |
| 1 | 0 | 1 | 1 | |
| 2 | 1 | 0 | 1 | $q^+ = q'$ toggle $q$ when $t = 1$ |
| 3 | 1 | 1 | 0 | |

With q as a table entered variable.

| row | t | $q^+$ | mode |
|-----|---|-------|------|
| 0 | 0 | q | hold |
| 1 | 1 | q' | toggle |

*Problem 617* Draw a schematic for a T flip-flop that uses a JK flip-flop.

*Problem 618* The next state equations for a set of logic functions are as follows. $q_0^+ = q_0\ xor\ 1$, $q_1^+ = q_1\ xor\ q_0'$, $q_2^+ = q_2\ xor\ q_1'q_0'$, $q_3^+ = q_3\ xor\ q_2'q_1'q_0'$. Design input circuits for T flip-flops.

*Problem 619* The truth table for a logic function is as follows. Design an input circuit for a T flip-flop.

| q | $q^+$ |
|---|-------|
| 0 | 1 |
| 1 | 0 |

*Problem 620* The truth table for a logic function is as follows. Design input circuits for T flip-flops.

| $q_1$ | $q_0$ | $q_1^+$ | $q_0^+$ |
|-------|-------|---------|---------|
| 0 | 0 | 0 | 1 |
| 0 | 1 | 1 | 0 |
| 1 | 0 | 1 | 1 |
| 1 | 1 | 0 | 0 |

## 6.4 Metastability

All flip-flops are clocked by the same clock signal in a synchronous state machine. Furthermore all of the waveforms at internal and output nodes are synchronous.

The problem of synchronization arises when an input to the system is not a synchronous waveform. For example, the input waveform frequency is different, or a step function is generated by a key press.

The problem is typically resolved by synchronizing the inputs with the system clock (5.3 page 99 Synchronizing Inputs, 5.5 page 102 State Assignment and Races).

However, this solution leads to violations of the setup and hold time window. When this is the case a flip-flop can enter a *metastable state*. Consequently the flip-flop's output is indefinite. It can switch as it is supposed to, it can go to an intermediate voltage level in the logic system forbidden band, it can oscillate for a while, and ultimately it can go to the correct or an incorrect level and remain there. Clearly a metastable state jeopardizes system reliability.

Metastable events can generate random failures in a digital system. In practice, metastable states are a source of disasters. There is no general solution to this problem[1][2][3]. All we can say is that experience is your guide.

---

[1] Chaney, T. C. 1983. *Measured Flip-flop Responses To Marginal Triggering.* IEEE Transactions on Computers C-32(12) (Dec): 1207-1209.

[2] `Kleeman, L., and A. Cantoni, 1987. *On the Unavoidability of Metastable Behavior in Digital Systems.* IEEE Transactions on Computers C-36(1) (Jan): 109-112.

[3] Johnson, H. 1993. *High-Speed Digital Design* pp132-143 ISBN 0133957242

# Review 6

***SR Latch*** A logic circuit acquires the memory property when a circuit output is connected to a circuit input. This connection from an output to an input is known as feedback. The most elementary logic circuit with the memory property is the SR Latch (set-reset latch).

*The SR latch defining equation is* $\quad q^+ = s + r'q$

***Asynchronous and Synchronous Circuits*** A logic circuit with no clock inputs is asynchronous. The outputs respond immediately to input changes. By immediately we mean after propagation delays elapse. An asynchronous circuit may or may not include feedback from one or more outputs to one or more inputs. I.e. an asynchronous circuit may or may not have memory. A logic circuit is synchronous when a clock is an input to all flip-flops in the circuit.

***Clock*** A clock is an independent signal source that emits a periodic signal. The transition from L to H is referred to as the positive edge, and the transition from H to L is referred to as the negative edge. The transition times are referred to as the rise and fall times respectively. The time between two consecutive positive, or two consecutive negative, transitions is the clock's period T that is the reciprocal of the clock's frequency f (T=1/f). A 500MHz clock has a 2ns period.

***The Setup and Hold Window*** A logic circuit with a clock whose positive or negative edges implement changes is a synchronous circuit. The selected clock edges (up or down but not both) are the active clock edges or triggering edges. Synchronous circuits have constraints similar to asynchronous circuits: only one input at a time should change. This means inputs should not change during a clock transition. These constraints are satisfied if no input changes inside the setup and hold time window. Only the clock should change inside the window.

***Edge Triggered Flip-Flop*** A logic circuit is a synchronous logic circuit when all bistable memory circuits are circuits whose outputs are changed only by active clock edges. A synchronous bistable memory circuit is called a flip-flop. It is an edge triggered flip-flop if a clock edge flips an output from L to H, and then some other clock edge flops it back to L. Other flip-flop inputs determine which clock edges flip and flop a flip-flop.

***D positive edge triggered flip-flop*** The $D_{FF}$ has a synchronous clock input and a data input d for defining the next state. Asserting the asynchronous preset or clear inputs sets or resets the q outputs, while overriding the synchronous inputs. While the asynchronous preset or clear inputs are not asserted, the defining equation is operational. The (present state) data at the d input meeting setup and hold requirements are transferred to the (next state) output q on the next positive clock edge. Triggering occurs when the edge rises up past a voltage level, and is not dependent on the edge's rise time over a wide range. After the hold time interval elapses data d may be changed with no effect on the outputs.

***JK positive edge triggered flip-flop*** The $JK_{FF}$ has a synchronous input clock, and data inputs j, k. The j and k inputs define the next state. Asserting the asynchronous preset or clear inputs sets or resets the q outputs, while overriding the synchronous inputs. While the asynchronous preset or clear inputs are not asserted, the defining equation is operational. The (present state) data at the j, k inputs meeting setup requirements are transferred to the (next state) output on the next positive clock edge. Triggering occurs when the edge rises up past a voltage level, and is not dependent on the edge's rise time over a wide range. After the hold time interval elapses data j, k may be changed with no effect on the outputs.

***T positive edge triggered flip-flop*** The T flip-flop is a special case of the jk flip-flop where t=j=k.

$$
\begin{aligned}
&\textit{Defining equations:} \\
&d \;\; \textit{flip-flop:} \quad q^+ = d \\
&jk \;\; \textit{flip-flop:} \quad q^+ = jq' + k'q \\
&t \;\; \textit{flip-flop:} \quad q^+ = tq' + t'q = t \; xor \; q
\end{aligned}
$$

***Metastability*** All flip-flops are clocked by the same clock signal in a synchronous state machine. Furthermore all of the waveforms at internal and output nodes are synchronous. The problem of synchronization arises when an input to the system is not a synchronous waveform. For example, a step function generated by a key press is asynchronous to the system. The problem is typically resolved by synchronizing the inputs with the system clock. However, this solution leads to violations of the setup and hold time window. When this is the case a flip-flop can enter a metastable state. There is no general solution to this problem. All we can say is that experience is your guide.

# 7 Complex Building Blocks with Memory

Complex building blocks with memory are state machines consisting of a register and logic gates. A *register* is a group of n flip-flops storing n bits that relate to each other in some way as defined by the wiring of the logic gates. The ASM method facilitates analysis and design of building blocks with memory such as counters, shift registers, and linear feedback shift registers

## 7.1 Writing data into a Register

*A register is a memory storing n bits* where each bit is stored in a D flip-flop. Data is written into the register's flip-flops by asserting an input called *load* (Figures 701, 702). Data in the register changes on a clock edge if load g is asserted. The defining equation is $q^+=gd+g'q$ (load d or load q). A two-to-1 multiplexer implements this equation (4.3.1 page 73). *(The standard register circuits are 4 bit 173, 8 bit 377)*

| **Figure 701 Load command action** | **Figure 702 D$_{FF}$ with load** |
|---|---|

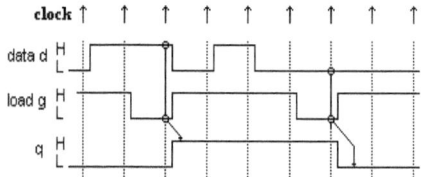
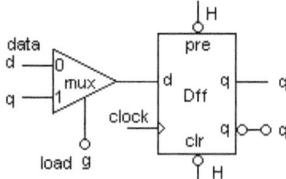

A one bit circuit (Figure 702) shows how a multiplexer controlled by load input g selects q (the present value) or d (new data) to be stored. Here are the defining equation and truth tables without and with table entered variables.

$$q^+ = gd + g'q \rightarrow$$

| $g$ | $d$ | $q$ | $q^+$ | action |
|---|---|---|---|---|
| 0 | – | 0 | 0 | store q |
| 0 | – | 1 | 1 | " |
| 1 | 0 | – | 0 | store d |
| 1 | 1 | – | 1 | " |

| $g$ | $g$ | $q^+$ | mux | action |
|---|---|---|---|---|
| H | 0 | q | 1 | store q |
| L | 1 | d | 0 | store d |

## 7.2 Counting Register

Counting registers report (store) the number of events received. Standard counting building blocks have up to four counter control lines: clear to zero (c), load data (g), count (p), count up or down (u). An ASM chart defines priorities and control line actions. We show how to use a flip-flop's *next state defining equation* to design a 3 bit up counter, and a 3 bit up/down counter.

A counter reports how many clock edge events occurred while count control p is true. Counters can count asynchronous events or synchronous events (e.g. count clock edges to measure time). Usually before counting starts the counter is reset to zero by the clear command c. Many applications require that the counter is set to some number N by the load command g. Then the counter may be required to count up or down from that preset number by the up/down command u. We start with a three-bit up counter.

### 7.2.1 Counter Control Lines

Counter registers are controlled by one to four variables. Operations are synchronous. That is not necessarily true of all commercial counters. Synchronous operations are vulnerable to glitches inside the setup and hold window (4.7.2 page 88, 6.2.2 page 113), whereas asynchronous operations are vulnerable to almost any glitch.

***Count p***: The number in the counter register is incremented by one (up counter) or decremented by one (down counter) at the next active clock edge when count p is asserted.

***Clear c***: The counter is reset to state $S_0$ at the next active clock edge after clear c is asserted. All bits are reset to zero.

***Load g***: The counter is set to state $S_N$ at the next active clock edge when load g is asserted and where N equals the base 2 number represented by the input data bits that are loaded.

***Up/down u***: The number in the register increments when $u \cdot p$ is asserted. When $u' \cdot p$ is asserted, the number in the register decrements (see Count p).

Priority issues arise when more than one control line is asserted in the same clock period. The ASM diamond tree (Figure 703) resolves priority issues.

Furthermore, the synchronous circuit requirement asserting *one input at a time* is satisfied when an ASM tree is implemented. The ASM diamond tree (Figure 703) allows more than one counter *input* to assert in the same clock period with no negative consequences, because only one tree *output* is asserted at any clock edge. The tree establishes priority automatically. For example, when the load function g is active, the count function p is disabled. The priority hierarchy used in standard building blocks is as follows.

| Input line | action | $c$ | $g$ | $p$ | $u$ | action | equation |
|---|---|---|---|---|---|---|---|
| $c$ | clear | 1 | – | – | – | clear to 00...0 | $c$ |
| $g$ | load | 0 | 1 | – | – | load count N | $c'g$ |
| $p$ | count | 0 | 0 | 0 | – | hold count n | $c'g'p'$ |
| $u$ | up / down | 0 | 0 | 1 | 0 | count down to $n-1$ | $c'g'pu'$ |
| | | 0 | 0 | 1 | 1 | count up to $n+1$ | $c'g'pu$ |

**Figure 703 ASM Diamond Trees for counter inputs c, g, p, and u**

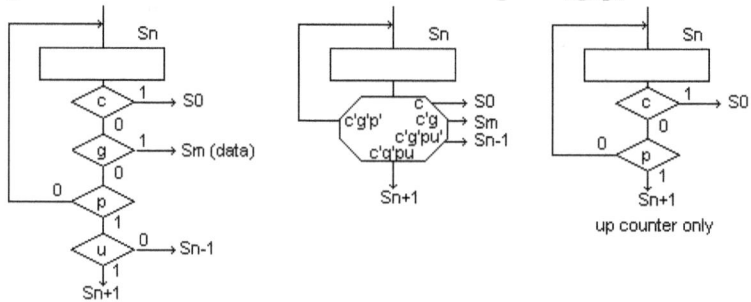

***Count p*** An XOR gate at input d of a $D_{FF}$ sets up toggle and hold.

$$q_k^+ = d_k = q \ xor \ p = p'q_k + pq_k'$$

$p = 0 \quad q_k^+ = d_k = q \ xor \ 0 = 1 \cdot q_k + 0 \cdot q_k' = q_k \quad (the \ hold \ condition)$

$p = 1 \quad q_k^+ = d_k = q \ xor \ 1 = 0 \cdot q_k + 1 \cdot q_k' = q_k' \quad (the \ toggle \ condition)$

## 7.2.2 Three-bit Synchronous Up Counter

The reasonable level of complexity of a three bit register allows us to illustrate in a straightforward way the *clear* and *count* features of a counter. The design is executed in 7.2.3 page 128. The present state is the count stored in the register configured as a counter. The present state, or count, is represented by the outputs $q_2q_1q_0$ of the register's three flip-flops.

In the decimal system a carry is added to the column to the left when all digits to the right are nines (1000=999+1). In the binary system a carry is added to the column to the left when all digits to the right are ones (10=1+1, 100=11+1, 1000=111+1).

An up counter accumulates ones 1+1+1+... The least significant bit $q_0$ of a counter counts 010101..., and each pass through 1 produces a carry to the next bit $q_1$. Each pass through $q_1q_0=11$ produces a carry to the next bit $q_2$, and so forth. The three bit counter state $q_2q_1q_0$ recycles from 111 to 000 as it counts up 0 to 7 to 0 to 7 to .... The $q_2q_1q_0$ binary strings show when each flip-flop changes state (Figure 704).

*(The standard 4 bit up counter with c, g, p controls is 163)*

| weight | output | binary  strings |
|---|---|---|
| 1 | $q_0$ | 0101 0101 0101 0101 0101.... |
| 2 | $q_1$ | 0011 0011 0011 0011 0011.... |
| 4 | $q_2$ | 0000 1111 0000 1111 0000.... |
| $state_{10}$ | $q_2q_1q_0$ | 0123 4567 0123 4567 0123.... |

**Figure 704 Up counter $q_2q_1q_0$ waveforms**

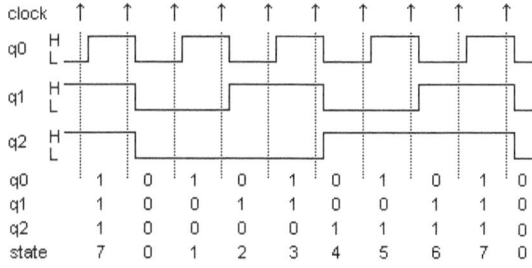

Digital Design

## 7.2.3 Design of a Three-bit Synchronous Up Counter

The ASM method is used to design a three bit up counter. Truth tables and equations are derived from the ASM chart (Figure 705). The results are

**Figure 705 Counter**

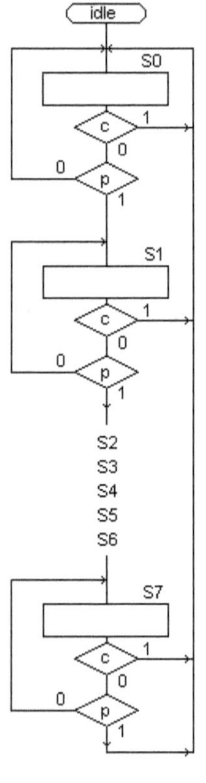

*Clear c=0* If count p= 0, then the counter holds the present count.

$$d_2 = q_2 xor\, 0 = q_2$$
$$d_1 = q_1 xor\, 0 = q_1$$
$$d_0 = q_0 xor\, 0 = q_0$$

*count p=1* If count p is 1, then the counter counts up.

$$d_2 = q_2 xor\, q_1 q_0 = q_2(q_1 q_0)' + q_2'(q_1 q_0)$$
$$\rightarrow d_2 = q_2'\ when\ q_1 q_0 = 1$$
$$d_1 = q_1 xor\, q_0 = q_1 q_0' + q_1' q_0 \rightarrow d_1 = q_1'\ when\ q_0 = 1$$
$$d_0 = q_0 xor\, 1 = q_0'$$

*c=0 and p=1* The ASM chart reduces to eight state rectangles without conditions.

| PS | NS | $q_2^+ q_1^+ q_0^+$ | |
|----|----|------|---|
| $s_0$ | $s_1$ | 001 | |
| $s_1$ | $s_2$ | 010 | |
| $s_2$ | $s_3$ | 011 | $d_0 = s_0 + s_2 + s_4 + s_6$ |
| $s_3$ | $s_4$ | 100 | $\Rightarrow d_1 = s_1 + s_2 + s_5 + s_6$ |
| $s_4$ | $s_5$ | 101 | $d_2 = s_3 + s_4 + s_5 + s_6$ |
| $s_5$ | $s_6$ | 110 | |
| $s_6$ | $s_7$ | 111 | |
| $s_7$ | $s_0$ | 000 | |

---

*Problem* 701 Use the up counter $d_2$, $d_1$, $d_0$ equations above to draw K maps with $q_2^+$, $q_1^+$, $q_0^+$ as coordinates. Then from the K maps derive new $d_2$, $d_1$, $d_0$ equations in xor format. Compare to the algebra on page 130.

---

***The Design*** The process of converting a *written* specification into an ASM flowchart is a task requiring creativity. Successful conversions require experience. We bypass the creative stage to start with an ASM flowchart with 8 states (Figure 705). This is an up counter. We implement the design process as follows:

**1.** Make Truth Table 701 listing the 8 states 000 to 111 in a present state column. Each present state is listed on two lines corresponding to the 2 values the variable p can have (0, 1). Assume c=0 to simplify the truth table.

**2.** Add present input p column to represent the next state logic equations. The expression in the diamonds is just *p*. In the present input p column list input values (0, 1) that cause exits to the next state.

**3.** Add a next state column. Given the present input values enter on each line the next state pointed to by the associated diamond exit.

**4.** Implement the state memory function with D flip-flops. Add columns for D flip-flop inputs $d_2$, $d_1$, $d_0$. Fill in the columns with the next state number.

**Table 701**

| Present State | p | Next State | $d2^+$ | $d1^+$ | $d0^+$ |
|---|---|---|---|---|---|
| S0 | 0 | S0 | 0 | 0 | 0 |
| S0 | 1 | S1 | 0 | 0 | 1 |
| S1 | 0 | S1 | 0 | 0 | 1 |
| S1 | 1 | S2 | 0 | 1 | 0 |
| S2 | 0 | S2 | 0 | 1 | 0 |
| S2 | 1 | S3 | 0 | 1 | 1 |
| S3 | 0 | S3 | 0 | 1 | 1 |
| S3 | 1 | S4 | 1 | 0 | 0 |
| S4 | 0 | S4 | 1 | 0 | 0 |
| S4 | 1 | S5 | 1 | 0 | 1 |
| S5 | 0 | S5 | 1 | 0 | 1 |
| S5 | 1 | S6 | 1 | 1 | 0 |
| S6 | 0 | S6 | 1 | 1 | 0 |
| S6 | 1 | S7 | 1 | 1 | 1 |
| S7 | 0 | S7 | 1 | 1 | 1 |
| S7 | 1 | S0 | 0 | 0 | 0 |

Digital Design

**5.** Derive the next state flip-flop input equations from Truth Table 701 (they require artistic xor manipulations, which are good experience).
The $d_0$ column in Table 701 has 8 ones, which means $d_0$ has 8 minterms.

$$d_0 = pm_0 + p'm_1 + pm_2 + p'm_3 + pm_4 + p'm_5 + pm_6 + p'm_7$$

$$= pq_2'q_1'q_0' + p'q_2'q_1'q_0 + pq_2'q_1q_0' + p'q_2'q_1q_0 + pq_2q_1'q_0' + p'q_2q_1'q_0 + pq_2q_1q_0' + p'q_2q_1q_0$$

$$= p(q_2'q_1'q_0' + q_2'q_1q_0' + q_2q_1'q_0' + q_2q_1q_0') + p'(q_2'q_1'q_0 + q_2'q_1q_0 + q_2q_1'q_0 + q_2q_1q_0)$$

$$= pq_0'(q_2'q_1' + q_2'q_1 + q_2q_1' + q_2q_1) + p'q_0(q_2'q_1' + q_2'q_1 + q_2q_1' + q_2q_1)$$

$$= pq_0' + p'q_0 = p \ xor \ q_0 \quad \rightarrow \quad d_0 = c'(q_0 \ xor \ p)$$

The $d_1$ column in Table 701 has 8 ones, which means $d_1$ has 8 terms. However they reduce to 6 minterms.

$$d_1 = pm_1 + (p' + p)m_2 + p'm_3 + pm_5 + (p' + p)m_6 + p'm_7$$

$$= pm_1 + m_2 + p'm_3 + pm_5 + m_6 + p'm_7$$

$$= pq_2'q_1'q_0 + q_2'q_1q_0' + p'q_2'q_1q_0 + pq_2q_1'q_0 + q_2q_1q_0' + p'q_2q_1q_0$$

$$= pq_2'q_1'q_0 + pq_2q_1'q_0 + p'q_2'q_1q_0 + p'q_2q_1q_0 + q_2'q_1q_0' + q_2q_1q_0'$$

$$= pq_1'q_0 + p'q_1q_0 + q_1q_0' = pq_1'q_0 + q_1(q_0' + p'q_0) = q_1'(pq_0) + q_1(q_0' + p')$$

$$= q_1'(pq_0) + q_1(pq_0)' = q_1 \ xor \ pq_0 \quad \rightarrow \quad d_1 = c'(q_1 \ xor \ pq_0)$$

The $d_2$ column in Table 701 has 8 ones, which means $d_2$ has 8 terms. However they reduce to 5 minterms.

$$d_2 = pm_3 + m_4 + m_5 + m_6 + p'm_7$$

$$= pq_2'q_1q_0 + q_2q_1'q_0' + q_2q_1'q_0 + q_2q_1q_0' + p'q_2q_1q_0$$

$$= pq_2'q_1q_0 + (p + p')q_2q_1'q_0' + (p + p')q_2q_1'q_0 + (p + p')q_2q_1q_0' + p'q_2q_1q_0$$

$$= q_2'(pq_1q_0) + q_2(q_1'q_0' + q_1'q_0 + q_1q_0' + p'q_1q_0)$$

$$= q_2'(pq_1q_0) + q_2(q_1'[q_0' + q_0] + q_1[q_0' + p'q_0])$$

$$= q_2'(pq_1q_0) + q_2(q_1' + q_1q_0' + p') = q_2'(pq_1q_0) + q_2(q_1' + q_0' + p')$$

$$= q_2'(pq_1q_0) + q_2(pq_1q_0)' = q_2 \ xor \ pq_1q_0 \quad \rightarrow \quad d_2 = c'[q_2 \ xor \ pq_1q_0]$$

$$used \ z' + zy = z' + y \ to \ get \ q_0' + p'q_0 = q_0' + p' \ and \ q_1' + q_1q_0' = q_1' + q_0'$$
$$and \ (q_1' + q_0' + p') = (pq_1q_0)' \ DeMorgan$$

***Discussion*** When in *any* state if clear c is true, then the next state is state $S_0$. When in any state if clear c is false and count p is true, then we have an up counter (the state transitions are 0 to 1, 1 to 2, 2 to 3, ... , 7 to 0). When count p is false the counter is stopped in the present state.

---

*Problem* 702 Synthesize logic circuits for the $d_2$, $d_1$, $d_0$ equations.

---

### 7.2.4 Three-bit Synchronous Up/Down Counter

A three-bit counter is has a 3-bit register whose present state is represented by the outputs $q_2q_1q_0$ of the register's three flip-flops. The present state is the current count stored in the counting register. A three-bit binary up counter can count from 0 to 7 on a recycling basis: 0, 1, 2, ... , 6, 7, 0, 1, 2, 3, etc. A down counter can count from 7 to 0 on a recycling basis. The outputs' binary strings show when each flip-flop changes state. The four counter control lines c, g, p, and u are used. (*The standard 4 bit up down counter with g, p, u controls is 169*)

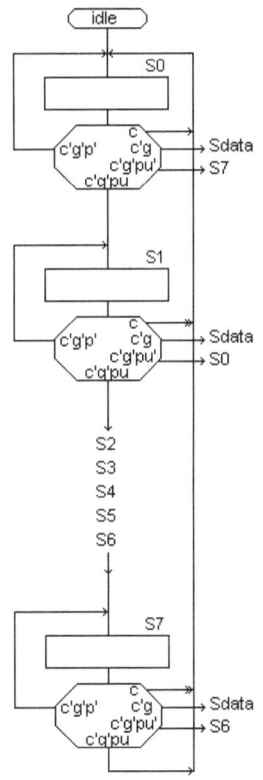

**Figure 707 Three bit counter state machine**

| weight | output | binary string |
|---|---|---|
| 1 | $q_0$ | 0101 0101 0101 0101 0101.... |
| 2 | $q_1$ | 0011 0011 0011 1001 1001.... |
| 4 | $q_2$ | 0000 1111 0000 0001 1110.... |
| $state_{10}$ | $q_2q_1q_0$ | 0123 4567 0123 2107 6543.... |
| $u$ | | 1111 1111 1110 0000 0000.... |

The three-bit counter's ASM chart has eight state rectangles, The output of each state rectangle drives a diamond tree (Figures 707, 703). Truth tables are derived from the ASM chart (Figure 707) as shown in Section 7.2.3 page 129. If clear c=1, then the next state is 000. If c=0 and count p=0 then the d flip-flops hold the present value $q_2q_1q_0$. $N_2$, $N_1$, $N_0$ are the data bits that can be loaded (Figure 708 page 132).

A counter is stopped, counts or loads. Loading data bits requires a c'gN$_k$ term.

$$d_2 = c'gN_2 + c'g'(q_2\, xor\, pq_1q_0) \quad \textit{count up}$$
$$d_1 = c'gN_1 + c'g'(q_1\, xor\, pq_0)$$
$$d_0 = c'gN_0 + c'g'(q_0\, xor\, p)$$

Up/down is more complicated. We leave the derivations to the problems. Note that $q_1q_0$ counts up so its $uq_1q_0$. Then we *guess* down is $u'q_1'q_0'$.

$$d_2 = c'gN_2 + c'g'[q_2 \ xor \ p(uq_1q_0 + u'q_1'q_0')]$$
$$d_1 = c'gN_1 + c'g'[q_1 \ xor \ p(uq_0 + u'q_0')]$$
$$d_0 = c'gN_0 + c'g[q_0 \ xor \ p]$$

**Figure 708 Three bit up/down counter circuit (clear c omitted)**

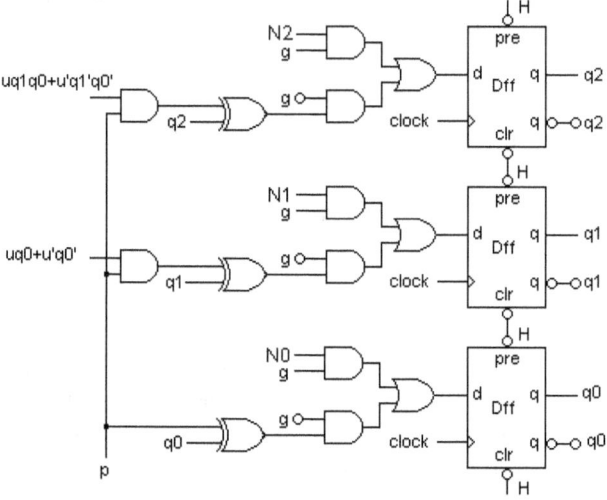

---

*Problem* 703 Use $q_1q_0$ as state variables. Make a truth table for a two bit synchronous down counter with present input count control p as a table entered variable. Draw the K maps for $q_1^+$ and $q_0^+$. Derive the $q_1^+$ and $q_0^+$ equations from the K maps:

*Problem* 704 Use $q_1q_0$ as state variables. Make a truth table for a two bit synchronous up/down counter with present input count control p and up/down control u as table entered variables. Draw the K maps for $q_1^+$ and $q_0^+$. Derive the $q_1^+$ and $q_0^+$ equations from the K maps:

*Problem* 705 Redraw Figure 708 showing clear control c. Hint. Merge the load g and clear c lines. Use gc' and g'c'.

*Problem* 706 Make K maps for a 3 bit up counter (Table 701 page 129). Map variables are $q_2^+$, $q_1^+$, $q_0^+$, and p as a table entered variable. Derive the $q_2^+$, $q_1^+$, $q_0^+$ equations from the maps.

---

## 7.2.5 Counters as Waveform Generators

Digital waveforms such as those in Figure 709 are either at the H level or at the L level. Time spent at either level is some multiple of the clock period when synchronous counters are used. The essential ideas applied to waveform generation are decoding of counter state numbers and presetting the counter to some state by the load command g, or to zero by the clear command c. The jk flip-flop is perfect for this application. The jk state transition Truth Table 702 is relevant here. For example, if the jk q is in state 0, then k is a don't care and j=1 in time slot m−1 produces a 0 to 1 output transition at the start of time slot m. The output waveform has NO glitches (Section 4.7 page 88).

| Truth Table 702 JK and D Flip-flop Transition Tables | | | |
|---|---|---|---|
| Transition | j | k | d |
| q from 0 to 0 | 0 | – | 0 |
| q from 0 to 1 | 1 | – | 1 |
| q from 1 to 0 | – | 1 | 0 |
| q from 1 to 1 | – | 0 | 1 |

**Figure 709 Digital Waveform**

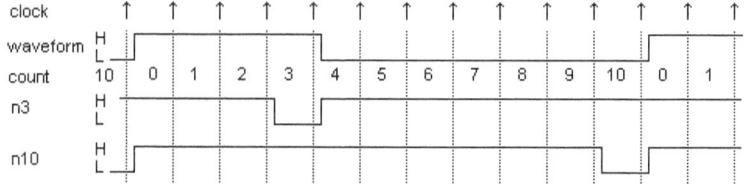

*A specific waveform* Suppose we want a waveform that has a period of 11 clock periods, and an output pulse that is 4 clock periods wide (*waveform* in Figure 709). This means the counter must count, from 0 to $10_{10}$, 0 to $10_{10}$, etc. In other words the counter is cleared, set to 0, every $11_{10}$ periods.

*Direct method* Implement the clear signal by decoding state $10_{10}$ (signal n10 Figure 709). Let the state ten decoder output $n_{10}$ assert load input g that will load 0000 at the next clock edge. Invert active low $n_{10}$ and assert the j input that will set the jk flip-flop output *waveform* to H at the next clock edge. Next decode state $3_{10}$. Let the state $3_{10}$ decoder active low output $n_3$ assert the k input that will set the *waveform* jk output to L at the next clock edge.

*Indirect method* This is illustrated by the upcoming CRT horizontal scan application.

***CRT monitor horizontal scan waveform*** A graphics display that is 1024 pixels wide requires 1024 clock periods to display 1024 pixels as the electron beam sweeps from left to right, and a somewhat arbitrary 192 clock periods to return the beam to the left for the start of the next line.

The *numbers 0 to 1023 (0 to 3FF$_{16}$) are used as memory addresses* to fetch the graphics bits to be displayed. The sweep is *blacked out* as it retraces from right to left as the counter outputs $h_0$ to $h_B$ (Figure 711) scan the 192 retrace clock periods. Therefore it does not matter what numbers are used by the counter during the retrace interval. The circuit design is simplified if we use numbers 3904 to 4095 (F40$_{16}$ to FFF$_{16}$) for the 192 retrace clock periods, because the FFF$_{16}$ decoder is available.

**Figure 710 Horizontal Scan Digital Waveforms**

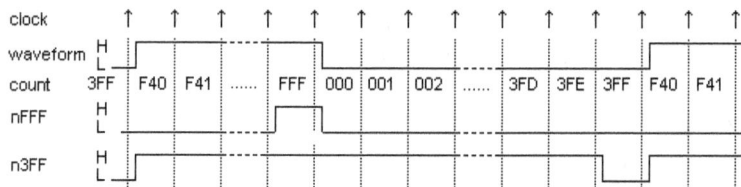

***CRT monitor horizontal scan circuit*** The 4 bit counter building block includes a state F$_{16}$ decoder. If we cascade three 4 bit counters, then the last state F$_{16}$ decoder is in fact a state FFF$_{16}$ decoder. This is why we use 3904 to 4095 (F40$_{16}$ to FFF$_{16}$) to generate the 192 clock period retrace (4095−3904+1=192). So, the scan across the screen starts with count equal to 0. When the count reaches 1023 (3FF$_{16}$) the decoded output n$_{3FF}$ is produced that loads 3904 (F40$_{16}$) into the counter. Then the counter proceeds to count from 3904 to 4095 (FFF$_{16}$) at which time decoded output n$_{FFF}$ is produced. Output n$_{FFF}$ could be used to load 000$_{16}$ into the counter or clear the counter. However no action is necessary because the next state is 000$_{16}$.

---

*Problem* 707 Reference Figure 710. Suppose this waveform represents 1111000000000. Design a circuit that produces the waveform. Use a 4 four bit counter and gates (also known as glue logic).

*Problem* 708 Design a down counter with state sequence FEDCBA98 using a four bit down counter and gates.

---

**Figure 711 Horizontal Scan Digital Circuit**

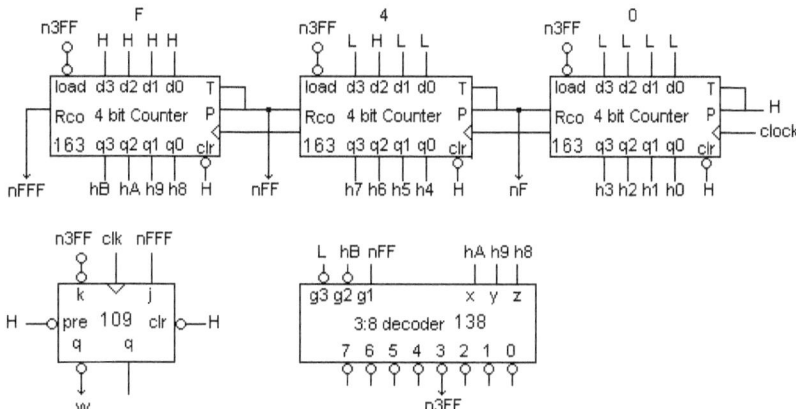

Since count 000 follows count FFF, $n_{FFF}$ does nothing to the counter. The counter proceeds to count up to 1023 ($3FF_{16}$) at which time decoded output $n_{3FF}$ is again produced to start the next waveform period. Decoded outputs $n_{3FF}$ and $n_{FFF}$ control a jk flip-flop, whose *active low q* generates the w *waveform*. Active high $n_{FFF}$ (j=1) is connected to the j input to change the *active high q* to H (w to L), and active low $n_{3FF}$ (k=1) is connected to the k input to change the *active high q* to L (w to H) (Figures 710, 711).

*Problem* 709 Design a 0 to $37_{10}$ up counter with states 0 to $36_{10}$ sequence using 4 bit counters and glue logic.

*Problem* 710 Design the circuit generating the waveform. Hint replace 0 to E with F1 to FF.

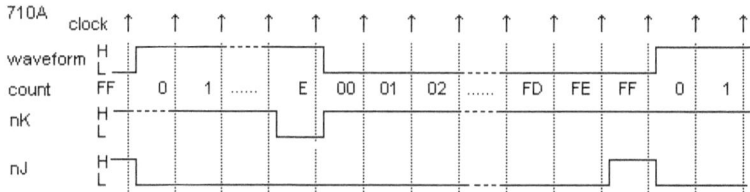

*Problem* 711 Design the circuit generating the waveform. Do not replace 0 to C with F1 to FF.

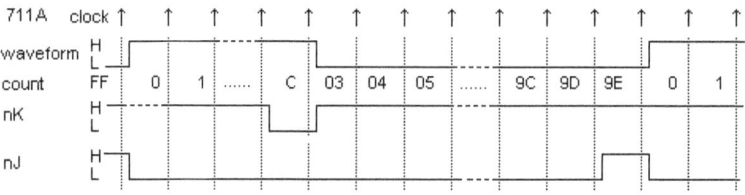

# 7.3 Shift Register

A *shift register* has the property that all bits stored in the register can move, on the same clock edge, one position to the left (or right) of their present position by asserting control lines. We design shift registers by applying the same methods used to design counters with emphasis on the *next state defining equation and ASM charts*.

The use of flip-flops as elements of a memory is relatively expensive when compared to cost per bit of a random access memory (RAM). The expense is justified by the *zero* access time. The flip-flop output nodes are directly accessible or accessible via a tri-state gate whose access time is say 2ns.

*Emphasis* The algorithmic state machine (ASM) method facilitates analysis and design of building blocks with memory, i.e. state machines, such as counters, and shift registers.

A *shift register* has the property that all bits stored in the register can move, on the same clock edge, one position to the left (or right) of their present position by asserting control lines. Here is the action in a 4-bit shift register. Note: *after* bits in positions 0 and 3 depend on the operation. Bit $q_L$ enters from the left on shift right. Bit $q_R$ enters from the right on shift left. The actions are valid for *any* number of shift register bits.

| code | action | before | after | equation |
|------|--------|--------|-------|----------|
| 00 | hold data | $q_3 q_2 q_1 q_0$ | $q_3 q_2 q_1 q_0$ | $d_k = d_k$ |
| 01 | right shift | $q_3 q_2 q_1 q_0$ | $q_L q_3 q_2 q_1$ | $d_k = d_{k+1}$ |
| 10 | left shift | $q_3 q_2 q_1 q_0$ | $q_2 q_1 q_0 q_R$ | $d_k = d_{k-1}$ |
| 11 | load data | $q_3 q_2 q_1 q_0$ | $q_{d3} q_{d2} q_{d1} q_{d0}$ | $d_k = q_{dk}$ |

| Operator | Bit pattern |
|----------|-------------|
| Before | $q_7\, q_6\, q_5\, q_4\, q_3\, q_2\, q_1\, q_0$ |
| | *After one shift* |
| ROTR | $q_0\, q_7\, q_6\, q_5\, q_4\, q_3\, q_2\, q_1$ |
| ROTL | $q_6\, q_5\, q_4\, q_3\, q_2\, q_1\, q_0\, q_7$ |
| SRL | $0\, q_7\, q_6\, q_5\, q_4\, q_3\, q_2\, q_1$ |
| SLL | $q_6\, q_5\, q_4\, q_3\, q_2\, q_1\, q_0\, 0$ |
| SRA | $q_7\, q_7\, q_6\, q_5\, q_4\, q_3\, q_2\, q_1$ |
| SLA | $q_6\, q_5\, q_4\, q_3\, q_2\, q_1\, q_0\, 0$ |

### 7.3.1 Shift Register Design Example

The right and left shift actions imply D flip-flop input equations $d_K = q_{K-1}$ for left shifts and $d_K = q_{K+1}$ for right shifts. Equations for the end flip-flop inputs

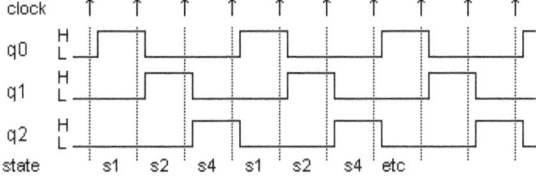

**Figure 712 3 Bit Ring Counter Waveforms**

depend on the operation to be implemented. Consider *an ad hoc design* of a circuit that produces *ring counter* waveforms (Figure 712).

The ring counter sequence of $q_2 q_1 q_0$ states is $s_1$, $s_2$, $s_4$, $s_1$, ....

The *waveforms* imply that the next state equations are $q_2^+ = q_1$, $q_1^+ = q_0$, $q_0^+ = q_2$. The equations imply that if power on state is 000 or 111, then the circuit will remain in $s_0$ or $s_7$ (000 or 111). We need to impose a *self correcting* condition by changing the equations so that 1 or 2 clock periods after power is turned on the sequence $s_1$, $s_2$, $s_4$ is entered.

The truth table shows that $q_1' q_0' = 1$ only in *present states* $s_0$, $s_4$, and that if $q_0^+ = q_1' q_0'$ the sequence $s_1$, $s_2$, $s_4$ is not changed. So change $q_0^+ = q_2$ to $q_0^+ = q_1' q_0'$ and evaluate the power-on properties of the new truth table.

**Building the Ring counter truth table**

*Seq is $s_1, s_2, s_4, s_1$....*          *change to $q_0^+ = q_1' q_0'$*

| PS | $q_2 q_1 q_0$ | $q_2^+ q_1^+ q_0^+$ | | PS | $q_2 q_1 q_0$ | $q_2^+ q_1^+ q_0^+$ | |
|----|------|------|---|----|------|------|---|
| $s_0$ | 000 | 001 | | $s_0$ | 000 | 001 | |
| $s_1$ | 001 | 010 | | $s_1$ | 001 | 010 | $q_2^+ = q_1$ |
| $s_2$ | 010 | 100 | | $s_2$ | 010 | 100 | |
| $s_3$ | 011 | | | $s_3$ | 011 | 110 | $q_1^+ = q_0$ |
| $s_4$ | 100 | 001 | | $s_4$ | 100 | 001 | |
| $s_5$ | 101 | | | $s_5$ | 101 | 010 | $q_0^+ = q_1' q_0'$ |
| $s_6$ | 110 | | | $s_6$ | 110 | 100 | |
| $s_7$ | 111 | | | $s_7$ | 111 | 110 | |

137

Digital Design

We derive the ASM chart (Figure 713) from the new truth table. The ASM chart shows that ring counter sequence $s_1$, $s_2$, $s_4$ is entered one or two clock periods after any power on state. The shift register ring counter circuit follows immediately from the q+ equations (Figure 714).

**Figure 713 Ring Counter ASM Chart**

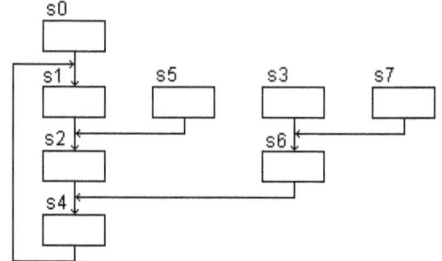

**Figure 714 Shift Register Ring counter**

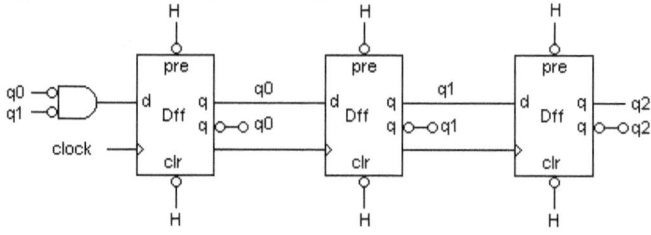

---

*Problem* 712 Draw the ASM chart for a 3-bit shift register implementing SRL.

*Problem* 713 Draw the ASM chart for a 4-bit shift register implementing ROTR.

*Problem* 714 Ring counter equations are $q_3{}^+=q_2$, $q_2{}^+=q_1$, $q_1{}^+=q_0$, $q_0{}^+=q_3$. Make a truth table. Draw the 16-state ASM chart.

*Problem* 715 Reference Figure 715. On power turn on assume the ring counter enters state $111_2$. What is the sequence of states leading into the correct sequence 001, 010, 001, 010, etc?

---

## 7.3.2 One Hots - State Numbers = $2^N$

State numbers equal to $2^N$ set *only one bit equal to 1* in the state register (e.g. for N = 0,1,2 the state numbers are 0001, 0010, 0100). The ring counter is one example of a one-hot state assignment. (Figure 712 p137). State numbers equal to $2^N$ are not encoded. The 3 bit shift register format is not encoded. The advantage is that some one-hot state machines may be faster, because encoding is not be required. The disadvantage is that more flip-flops are required. For example consider the one hot 3 bit ring counter.

| PS | $q_2 q_1 q_0$ | $q_2^+ q_1^+ q_0^+$ | NS |
|---|---|---|---|
| $s_0$ | 000 | 001 | $s_1$ |
| $s_1$ | 001 | 010 | $s_2$ |
| $s_2$ | 010 | 100 | $s_4$ |
| $s_3$ | 011 | 000 | $s_0$ |
| $s_4$ | 100 | 001 | $s_1$ |
| $s_5$ | 101 | 000 | $s_0$ |
| $s_6$ | 110 | 000 | $s_0$ |
| $s_7$ | 111 | 000 | $s_0$ |

$$q_0^+ = d_0 = m_0 + m_4$$
$$q_1^+ = d_1 = m_1$$
$$q_2^+ = d_2 = m_2$$

A 3:8 decoder generates the minterms of $q_2 q_1 q_0$ (Figure 715). If term $q_2 q_1 q_0$ is in state k, then output k ($m_k$) is L.

Therefore if k = 0 or 4 only $m_0$ or $m_4$ are L. Then the $q_2 q_1 q_0$ are set to LLH. The next state is 001.

And, if k = 1 or 2, then only $q_1$ or $q_2$ is set to H (states 010, 100).

**Figure 715 A Different Ring Counter**

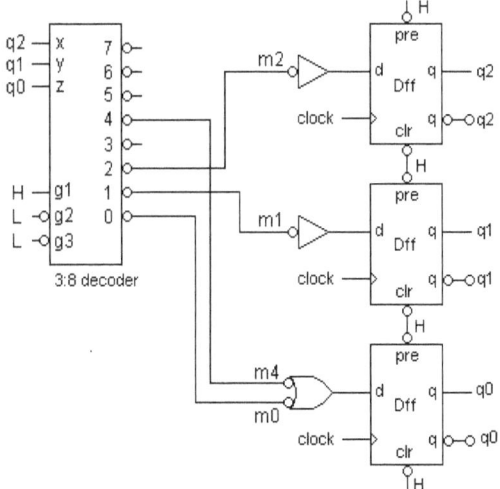

If the circuit is in state 3, 5, 6, or 7 when it turns on, then the next state is $S_0$ (000), because minterms 0, 1, 2, 4 are H forcing the d inputs L. Once the circuit is in state 0, the next state is 1. This starts the ring counter.

# 7.4 Linear Feedback Shift Register (LFSR)

A Linear Feedback Shift Register (LFSR) is an n-bit counter that has a *pseudo-random* state sequence. LFSR design requires knowledge of Galois field theory that we introduce and point the way to further study. We do this to illustrate how sophisticated mathematics is applied to digital circuit design. There is no need to hack the LFSR design process.

The synchronous linear feedback shift register (LFSR) is an assembly of D flip-flops and one or more *feedback* XOR gates in a shift-register structure. In a *Galois* LFSR the XOR gates are inserted between the D flip-flops (Figure 716). The clock and reset lines are not shown. In a Galois m bit LFSR the bits are numbered 0 to m−1. A feedback line from $bit_{m-1}$ is connected to one input of each XOR gate. A bit output is connected to the other input of each XOR gate. An exception is the $bit_0$ XOR gate that is connected to $bit_0$'s input. There, the serial input to the LFSR is connected to the other $bit_0$ XOR input (Figure 716). The circuit is derived from a *Galois primitive polynomial $x^3+x+1$* (7.4.1).

**Figure 716 Three bit Linear Feedback Shift Register - Galois $x^3+x+1$**

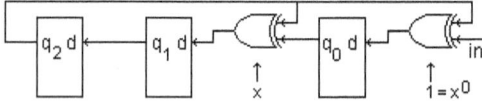

The number of $D_{ff}$ equals the degree of the primitive polynomial. The feedback originates at the leftmost $D_{ff}$ and is fed back to all $D_{ff}$ marked by powers of x in the polynomial.

The LFSR that has the Galois form (Figure 716) that is much faster than the *Fibonacci* form whose xor gates form a tree of xor gates (Figure 717).

**Figure 717 Three bit Linear Feedback Shift Register - Fibonacci $x^3+x+1$**

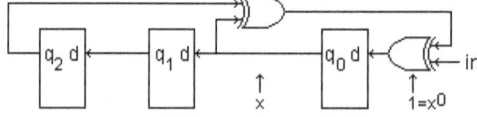

*Problem* 716 Reference Figure 716. Connect the input to L. Initialize the LFSR to 001. Show that the sequence of states is 001, 010, 100, 011, 110, 111, 101, 001, 010, etc.

## 7.4.1 From Primitive Polynomial to LFSR

Someone discovered that *primitive polynomials are LFSR design equations.* We have not been able find out who. Therefore we cannot tell you their important reasoning that led to the connection, nor give credit to the person(s).

The mathematical theory of LFSRs is derived from E. Galois' (1811-1832) finite field theory $GF(2^m)$, and K. F. Gauss' (1777-1855) concept of congruence applied to operations $+ - \times \div$ on polynomials so that binary strings (k-tuples) can be manipulated by interpreting their 0's and 1's as coefficients of terms of polynomials.

An LFSR will sequence through all states in some pseudo random way if the polynomial representation is *primitive.* A *primitive* polynomial f(x) is an irreducible polynomial that is a factor of $x^n + 1 = 0$ *where* $n = 2^m - 1$ (Polynomials Mod 2 page 144). Primitive polynomials are factors of $GF(2^m) = x^n + 1$ *where* $n = 2^m - 1$

$$GF(4) = x^3 + 1 = (x+1)(x^2 + x + 1)$$
$$GF(8) = x^7 + 1 = (x+1)(x^3 + x + 1)(x^3 + x^2 + 1)$$
$$GF(16) = x^{15} + 1 = (x+1)(x^2 + x + 1)(x^4 + x^3 + x^2 + x + 1)(x^4 + x^3 + 1)(x^4 + x + 1)$$

The 3 bit Galois LFSR design equation is *primitive polynomial $x^3 + x + 1$* (Figure 716). The 3 bits have $n = 2^3 - 1 = 7$ states when the all-zeros state is excluded. The next LFSR state (Figure 716) is calculated in the usual way.

$$q_3^+ = q_2 \qquad q_2^+ = q_3 \oplus q_1 \qquad q_1^+ = q_3 \oplus q_{in}$$

*A complete explanation of this very complex theory is outside the scope of this text.* Nevertheless we explain some statements.

---

**Primitive Polynomials of GF($2^m$), m=2, 3, 4**

| | |
|---|---|
| $x^2 + x + 1$ | $x^4 + x + 1$ |
| $x^3 + x + 1$ | $x^4 + x^3 + 1$ |
| $x^3 + x^2 + 1$ | $x^4 + x^3 + x^2 + x + 1$ |

---

## 7.4.2 Counter and Pseudo Random Number Generator

A linear-feedback shift register (LFSR) is one alternative to a conventional binary counter in some applications, because the LFSR counter count sequence is not sequential.

Any counter specification includes a maximum number n, and any m flip-flop state machine has $2^m-1$ non-zero states. Consequently n is less than or equal to $2^m-1$ so that $m_{min} \geq \log_2(n+1)$. For example, a divide by 47 counter requires m = 6 flip-flops [ $6 \geq \log_2(47+1)$ *so that* $2^6 = 64 > 48$ ].

The all-zeros state is not allowed in an LFSR state machine, because the LFSR may remain in that state (a lock up). This is why a reset circuit is required to set the counter to a non zero state at power-on startup or other conditions.

Question: After k clocks are applied to the LFSR what is the LFSR state? We have not found any theory that answers this question.

***Example*** Division by k is one major use of an LFSR counter. Here is how an LFSR counter that divides by k is implemented. Assume a divide-by-12 counter is required. The smallest power of 2 greater than 12 is $2^4$. From *Primitive Polynomials of GF($2^m$)* (page 141) select the fourth-degree polynomial $x^4+x+1$. This circuit requires 4 flip-flops, and two XORs (Figure 718). For the divide by 12 reset circuit see Problem 718.

**Figure 718 Four bit Linear Feedback Shift Register - Galois $x^4+x+1$**

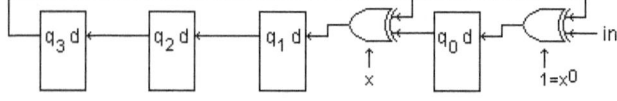

---

*Problem* 717 Reference Figure 718. Connect the input to L. Initialize the LFSR to 0001. Show that the sequence of states is 1248 36CB 5A7E FD91... etc.

*Problem* 718 Reference Figure 718. The sequence of states is 1248 36CB 5A7E FD91... etc. The twelfth state is E=1110 that is decoded to reset the LFSR to 1=0001. Add the divide by 12 reset circuit to Figure 718.

---

### 7.4.3 Polynomial Division

LFSRs divide input polynomials by a primitive polynomial. An LFSR defined by primitive polynomial $p(x)$ divides *input h(x) by p(x)* to produce remainder $r(x)$ stored in the LFSR flip-flops, where $h(x)=q(x)p(x)+r(x)$.

*Long division* Divide polynomial $h(x)=x^8+x^5+x^4+x^2+1$ (100110101) by irreducible primitive polynomial $p(x)=x^3+x+1$ (1011) to get remainder $r(x)=x^2+x$ (110). In the sidebar *Polynomials Mod 2* we will show an easy way to divide by a primitive polynomial.

$$
\begin{array}{r}
1x^5+0x^4+1x^3+0x^2+0x^1+1x^0 \quad 101001 \\
\hline
x^3+x+1\overline{\smash{\big)}\,1x^8+0x^7+0x^6+1x^5+1x^4+0x^3+1x^2+0x^1+1x^0}
\end{array}
$$

$$
\begin{array}{l}
1x^8 \qquad\quad +1x^6+1x^5 \\
\hline
\quad 0x^7+1x^6+0x^5+1x^4 \qquad\qquad\qquad 010 \\
\quad 0x^7 \qquad\quad +0x^5+0x^4 \\
\hline
\qquad 1x^6+0x^5+1x^4+0x^3 \qquad\qquad\quad 101 \\
\qquad 1x^6 \qquad\quad +1x^4+1x^3 \\
\hline
\qquad\quad 0x^5+0x^4+1x^3+1x^2 \qquad\qquad 001 \\
\qquad\quad 0x^5 \qquad\quad +0x^3+0x^2 \\
\hline
\qquad\qquad 0x^4+1x^3+1x^2+0x^1 \qquad\quad 011 \\
\qquad\qquad 0x^4 \qquad\quad +0x^2+0x^1 \\
\hline
\qquad\qquad\quad 1x^3+1x^2+0x^1+1x^0 \qquad 110 \\
\qquad\qquad\quad 1x^3 \qquad\quad +1x^1+1x^0 \\
\hline
\qquad\qquad\qquad 1x^2+1x^1+0x^0 \qquad 110
\end{array}
$$

**Figure 716 Three bit Linear Feedback Shift Register - Galois**

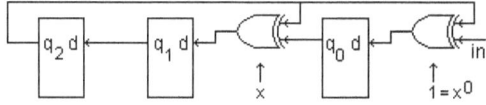

If the bit stream 100110101 is fed into the LFSR (Figure 716) the last LFSR state is the remainder 110 (Problem 719).

---

*Problem* 719 Reference Figure 716 ($x^3+x+1$). Initial LFSR state is 000. Input bit stream is 10011010 ($x^8+x^5+x^4+x^2+1$), Show that sequence of states is 001, 010, 100, 010, 101, 001, 011, 110, 110, where 110 is the remainder. Also show that the $q_2$ sequence is 00101001 (the quotient).

---

## Polynomials Mod 2

LFSR's divide an input k-tuple (a k digit binary string) by the LFSR defining n-tuple. This is one example of the need to manipulate binary k-tuples. Coding theory is another application where we need to manipulate binary k-tuples (8.4 page 176).

*The binary k-tuple 100110101 may be interpreted as a polynomial*

$$f(x) = 1x^8 + 0x^7 + 0x^6 + 1x^5 + 1x^4 + 0x^3 + 1x^2 + 0x^1 + 1x^0$$
$$f(x) = x^8 + x^5 + x^4 + x^2 + 1$$

When we add, multiply, and divide polynomials, the operations on the binary coefficients are executed modulo 2 so that all resulting coefficients are 1 or 0. The zero k-tuple (0000...00) corresponds to the zero

$$x^5 + x^4 + x^3 + x^2 + x^1$$
$$+x^5 \qquad\qquad +x^2 + x^1 + 1$$
$$\overline{\qquad x^4 + x^3 \qquad\qquad +1}$$

polynomial. The sum of any two polynomials is another polynomial of the same degree or less. Addition modulo 2 on coefficients implement 0+0=0, 0+1=1, 1+0=1, and 1+1=0. (Subtraction operations produce the same results as addition operations.) Emphasis: $x^r + x^r = 1x^r + 1x^r = (1+1)x^r = 0x^r = 0$.

$$x^3 + x^2 + 1$$
$$\underline{\times \quad x^3 + x^2 + x^1}$$
$$x^4 + x^3 + \qquad x^1$$
$$x^5 + x^4 \qquad + x^2$$
$$\underline{+ \ x^6 + x^5 \qquad + x^3}$$
$$x^6 \qquad\qquad + x^2 + x^1$$

```
     1101
   ×1110
     0000
     1101
     1101
  +1101
  1000110
```

The degree of a product of two polynomials is the sum of the degrees of the two polynomials. For example calculate the product of two polynomials.

*In a mod k system the polynomial degree is limited to the range 0 to k−1. This means the degree of a polynomial product has to be reduced modulo by some polynomial of degree k. Furthermore the divider must be primitive.*

Polynomials whose degree is greater than k−1 are divided by a *primitive* polynomial of degree k and the remainder with degree ≤ k−1 replaces the polynomial divided.

We can do division in two ways. We can use the polynomials directly, or we can replace the polynomials with binary numbers derived from the coefficients to simplify the writing of the division process. For example:

$$x^4 + x^3 + 1 \overline{\smash{)}\, x^6 + x^2 + x} \qquad becomes \qquad 11001\overline{\smash{)}\,1000110}$$

$$
\begin{array}{r}
111 \\
11001\overline{\smash{)}\,1000110} \\
11001 \\
\hline
100010 \\
11001 \\
\hline
10000 \\
11001 \\
\hline
1001
\end{array}
$$

$$h(x) = q(x)p(x) + r(x)$$

$$x^6 + x^2 + x = (x^2 + x^1 + 1)(x^4 + x^3 + 1) + (x^3 + 1)$$

$$1000110 = 111 \times 11001 + 1001$$

Since multiplication of polynomials f(x) with degree f, and g(x) with degree g results in a polynomial h(x) with degree f+g > k, the binary coefficients of this polynomial cannot be interpreted as a binary k-tuple. This why we have divided h(x) by some primitive p(x) with degree k to produce remainder r(x) with degree $r \le k-1$.

$$h(x) = f(x)g(x)$$

$$\frac{h(x)}{p(x)} = \frac{f(x)g(x)}{p(x)} = q(x) + \frac{r(x)}{p(x)}$$

$$h(x) = q(x)p(x) + r(x) \qquad [\deg r(x) < \deg p(x) < \deg h(x)]$$

$$h(x) = r(x) \ mod \ p(x) \qquad p(x) \ primitive$$

$$this \ means \ h(x) - r(x) \ is \ divisible \ by \ p(x)$$

*P(x) must be primitive.* Primitive means p(x) has *no factors with binary coefficients, and is a factor of $x^n - 1$ where $n = 2^m - 1$.* For example.

$$x^{2^3 - 1} + 1 = x^7 + 1 = (x+1)(x^3 + x + 1)(x^3 + x^2 + 1)$$

$$x^{2^4 - 1} + 1 = x^{15} + 1 = (x+1)(x^2 + x + 1)(x^4 + x^3 + x^2 + x + 1)(x^4 + x^3 + 1)(x^4 + x + 1)$$

This what we have been leading up to -- *An easy way to manipulate the polynomials h(x)=q(x)p(x)+r(x)*

*If $\alpha$ is a root, then divisor $p(\alpha) = 0$ and*

$$h(\alpha) = q(\alpha)p(\alpha) + r(\alpha) = q(\alpha) \cdot 0 + r(\alpha) = r(\alpha)$$

*or*

$$h(\alpha) = r(\alpha) \ mod \ p(\alpha) \qquad this \ means \ h(\alpha) - r(\alpha) \ is \ divisible \ by \ p(\alpha)$$

We have shown (7.4.3 page 143) that

$$h(x) = x^8 + x^5 + x^4 + x^2 + 1 = x^2 + x \ \ mod \ \ x^3 + x + 1$$

*and we show below that $j(x) = x^6 + x^2 + x = x^3 + 1 \ \ mod \ \ x^4 + x^3 + 1$*

*If $\alpha$ is a root of primitive polynomial $p(x)=x^3+x+1$, then*

$$0 = \alpha^3 + \alpha + 1 \ \rightarrow \ (\alpha + 1) = \alpha^3 + \alpha + 1 + (\alpha + 1) = \alpha^3$$

$$\alpha^3 = \alpha + 1 \ \ mod \ 2 \qquad because \qquad \alpha + 1 + \alpha + 1 = 0 \ \ mod \ 2$$

Watch this – the hidden logic here is that using $\alpha$ means dividing by p(x).

$$h(\alpha) = \alpha^8 + \alpha^5 + \alpha^4 + \alpha^2 + 1$$

$$= \alpha^3 \alpha^3 \alpha^2 + \alpha^3 \alpha^2 + \alpha^3 \alpha + \alpha^2 + 1$$

$$= (\alpha + 1)^2 \alpha^2 + (\alpha + 1)\alpha^2 + (\alpha + 1)\alpha + \alpha^2 + 1$$

$$= (\alpha^2 + 1)\alpha^2 + \alpha^3 + \alpha^2 + \alpha^2 + \alpha + \alpha^2 + 1$$

$$= (\alpha + 1)\alpha + (\alpha + 1) + \alpha + 1 = \alpha^2 + \alpha$$

$$h(x) = x^2 + x \ \ mod \ \ x^3 + x + 1$$

And j(x) simplifies as follows where $p(\alpha)=\alpha^4+\alpha^3+1=0$ as $\alpha^4=\alpha^3+1$

$$j(\alpha) = \alpha^6 + \alpha^2 + \alpha = \alpha^4 \alpha^2 + \alpha^2 + \alpha$$

$$= (\alpha^3 + 1)\alpha^2 + \alpha^2 + \alpha \ \rightarrow \ \alpha^4 = \alpha^3 + 1$$

$$= \alpha^5 + \alpha^2 + \alpha^2 + \alpha = \alpha^5 + 0 + \alpha = \alpha^4 \alpha + \alpha = (\alpha^3 + 1)\alpha + \alpha$$

$$= \alpha^4 + \alpha + \alpha = \alpha^3 + 1 + 0$$

$$= \alpha^3 + 1 \ \ mod \ \alpha^4 + \alpha^3 + 1$$

$$j(x) = x^3 + 1 \ \ mod \ \ x^4 + x^3 + 1$$

We think of the above as

| This is Galois magic. |
|---|

# 7.5 Euclid's gcd Calculator

The gcd calculation can be broken down into three parts – entering the digits of numbers M and N, converting the digits to binary numbers M and N, and calculating the gcd of M and N.

The mathematical processes calculating the gcd and converting digits to numbers specified two parts of the ASM (Section 5.6 page 104 and Figure 719 page 148). The third part, entering the digits of numbers M and N, was just a matter of saying how to process key presses.

Turning on the power clears all storage bits to 0 and sets the ASM to state $S_0$. A reset button does the same. The M, N and gcd displays show all zeros. With the ASM in state $S_0$, variable gcd, or M, or N is activated by pressing one of three front panel push buttons moving the ASM to $S_5$ or $S_1$.

We want to use the same hardware to fetch M and N digits. Therefore we show we are fetching M or N digits by setting one bit storage flip-flops Md to 1 or Nd to 1, which will direct digits to M or N digit storage.

Next we press M and then N to enter and store the digits of M and N.

Now we can calculate the gcd of M and N by pressing gcd, which moves the ASM to $S_5$. $S_5$ activates a linked ASM that converts M and N digits to numbers M and N (Figures 721, 722 page 149). When done, the ASM moves to $S_4$ to calculate the gcd (Figure 719 page 148).

In $S_4$ the first subtraction takes place (page 106). Repeated subtractions occur until R=0 (and N is the gcd) or R<N, which moves the ASM to $S_6$ where N is stored in M. The ASM moves to $S_7$ where R (the last M–N) is stored in N. and the cycle of repeated subtractions repeats until R=0 or R<N again. The process ends when R=0.

We move on to develop the circuits storing digits, the circuits converting digits to numbers, and the circuits calculating the gcd. .

The ASM truth table and equations are derived, which allows us to design the control and the datapath.

## Figure 719 Euclid's gcd ASM

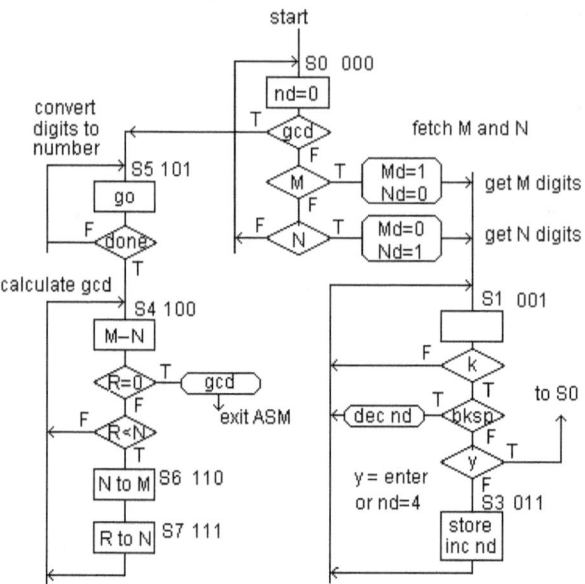

| The Convert truth table | | | | |
|---|---|---|---|---|
| $q_1q_0$ | Go | nd=4 | MNd | $q_1^+q_0^+$ |
| 00 | 0 | - | - | 00 |
| 00 | 1 | - | - | 01 |
| 01 | - | - | - | 11 |
| 10 | - | - | - | 01 |
| 11 | - | 0 | 0 | 01 |
| 11 | - | 0 | 1 | 01 |
| 11 | - | 1 | 0 | 10 |
| 11 | - | 1 | 1 | exit |

The M, and N digits are stored in flip-flops whose outputs drive LED 7 segment displays (Figure 720). Tri-state gates allow digits to be read out onto a bus for subsequent processing into M and N numbers.

## Figure 720 Fetch Digits Datapath for M or N

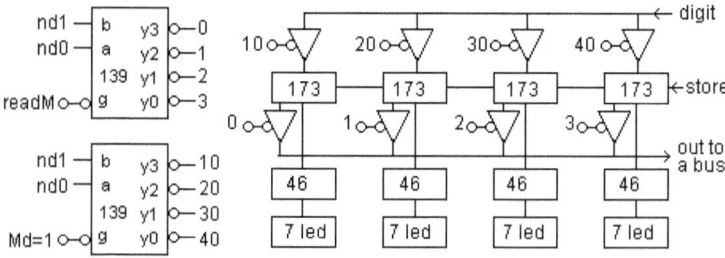

**Figure 721 ASM₂ Convert Digits**

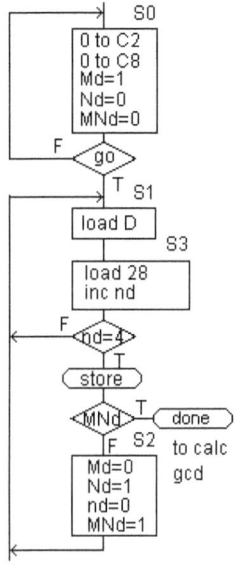

This link ASM₂ (Figure 721) is linked to the main ASM state S₅ (Figure 719). When ASM₂ receives a *go* it moves to S₁ where the most significant M digit is stored by *load d* in register D (Figure 722) and ASM₂ moves to S₃.

Assuming the M digits are 6, 7, 8, 9 the 6 is now in D. Since registers $C_2$ and $C_8$ are cleared to zero the add₂ sum is 6 (Figure 722). In S₃ a *load 28* stores 2×6 in $C_2$ and 8×6 in $C_8$ so that *add1 out* is 60. And, digit number nd is increased from 0 to 1.

Since nd<4 the next state is S₁ where digit 7 is stored in D so that *add2 out* is 67. The process repeats until add₂ *out* is the number 6,789 and nd=4. Then 6,789 is stored in register M, because Md=1 (Figure 719 S₀), and the ASM moves to S₂, because MNd=0 (F). In S₂ the circuit is set up to convert digits N.

**Figure 722 Convert Datapath**

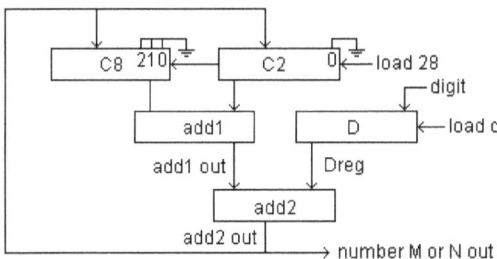

The converted from digits numbers M and N are stored in registers M and N (by load N and load M) as shown in Figure 723 page 150. We are back in the main ASM at S₄. To create M minus N an inverter complements N, 1 is added by $c_{in}$, and the sum is added to M. M minus N is compared to N and to 0 to get outputs R>N and R=0. If R>0 and R>N the ASM moves to S₄ where M minus N is stored in M to allow for the next subtraction of N. This process proceeds until R<N and the ASM moves to S₆ to set up the next round of subtractions (S₆ stores N in M and S₇ stores R in N). When R = 0 the gcd is stored in register G (Figure 723).

## Figure 723 Calculate gcd Datapath

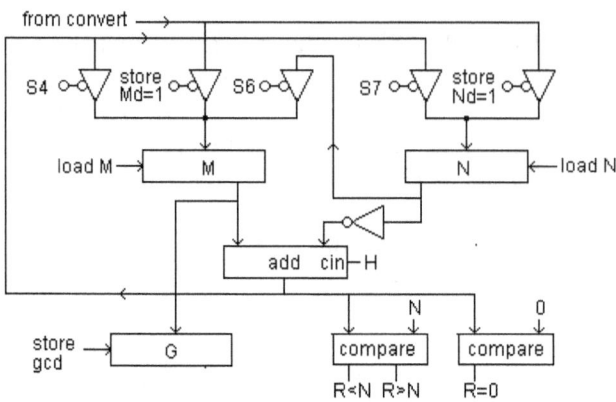

| Truth Table 703 Euclid's ASM (Figure 719) | | | | |
|---|---|---|---|---|
| $q_2q_1q_0$ | gcd | M | N | $q_2^+q_1^+q_0^+$ |
| 000 | 0 | 0 | 0 | 000 |
| 000 | 0 | 0 | 1 | 001 |
| 000 | 0 | 1 | – | 001 |
| 000 | 0 | 1 | – | 001 |
| 000 | 1 | – | – | 101 |
| | | | | |
| | k | bksp | Y | |
| 001 | 0 | – | – | 001 |
| 001 | 1 | 0 | 0 | 011 |
| 001 | 1 | 0 | 1 | 000 |
| 001 | 1 | 1 | – | 001 |
| | | | | |
| 011 | – | – | – | 001 |
| | | | | |
| | R=0 | R<N | | |
| 100 | 0 | 0 | – | 100 |
| 100 | 0 | 1 | – | 110 |
| 100 | 1 | 0 | – | Exit |
| | | | | |
| | done | | | |
| 101 | 0 | | | 101 |
| 101 | 1 | | | 100 |
| | | | | |
| 110 | – | – | – | 111 |
| 111 | – | – | – | 100 |

*Problem* 722 Design the ASM control circuit using mux, $D_{FF}$, and the $d_2$, $d_1$, $d_0$ equations.

*Problem* 723 Convert Figures 720, 722, 723 into xyz bus format as shown in Figure 11 page 15.

*Problem* 724 The gcd number is in binary format. How would you convert it to four digits that you can display using a Figure 720 circuit?

The control D flip-flop input equations are taken from Truth Table 703.

$$d_2 = \text{gcd} \cdot S_0 + (R=0)' \cdot S_4 + S_5 + S_6 + S_7$$
$$d_1 = k \cdot bksp' \cdot enter' \cdot S_1 + (R=0)' \cdot (R<N) \cdot S_4 + S_6$$
$$d_0 = \text{gcd}' \cdot M' \cdot N \cdot S_0 + \text{gcd}' \cdot M \cdot S_0 + \text{gcd} \cdot S_0 + k' \cdot S_1 + k \cdot bksp' \cdot enter' \cdot S_1$$
$$\qquad + k \cdot bksp \cdot S_1 + S_3 + done' \cdot S_5 + S_6$$
$$d_0 = \text{gcd}' \cdot (M'N + M) \cdot S_0 + \text{gcd} \cdot S_0$$
$$\qquad + (k' + k \cdot bksp' \cdot enter') \cdot S_1 + k \cdot bksp \cdot S_1$$
$$\qquad + S_3 + done' \cdot S_5 + S_6 \qquad\qquad next \quad we \; use \; x + x'y = x + y$$
$$d_0 = [\text{gcd}' \cdot (N + M) + \text{gcd}] \cdot S_0 + (k' + bksp' \cdot enter' + k \cdot bksp) \cdot S_1$$
$$\qquad + S_3 + done' \cdot S_5 + S_6$$
$$d_0 = (N + M + \text{gcd}) \cdot S_0 + (k' + enter' + bksp) \cdot S_1 + S_3 + done' \cdot S_5 + S_6$$

*Output equations* We will not create an output truth table, because it is just easier to write the equations by examining the ASM in Figure 719..

$$(n=0) = S_0$$

| | |
|---|---|
| $(Md = 1) = \text{gcd}' \cdot M \cdot S_0$ | $(Nd = 0) = \text{gcd}' \cdot M \cdot S_0$ |
| $(Md = 0) = \text{gcd}' \cdot M' \cdot N \cdot S_0$ | $(Nd = 1) = \text{gcd}' \cdot M' \cdot N \cdot S_0$ |

$$dec \; nd = k \cdot bksp \cdot S_1$$
$$store = S_3 \qquad inc \; nd = S_3$$

| | | |
|---|---|---|
| $go = S_5$ | $M - N = S_4$ | $\text{gcd} = (R=0) \cdot S_4$ |

$$NtoM = S_6$$
$$RtoN = S_7$$

# Review 7

*Writing data into a Register* The D flip-flop defining equation, $q^+ = d$, shows that at each active clock edge a D flip-flop stores the present value of d that becomes the next value of q. Data is stored, however, with the intention of *saving* the data for future use that means data is stored on selected clock edges. Thus a register cannot be just a group of D flip-flops. Something else is needed. The act of storing data is called loading. This implies an input line named *load* that controls a mux that stores new data or holds present data. A storage register is an assembly of D flip-flops with the load function.

*Counting Register* Synchronous look-ahead carry up and down counters are introduced. We start with a three-bit up counter so that we can focus on basic principles outside the shadows of the many details associated with larger numbers of bits.

Controls clear c, load g, count p, and up/down u are defined. A three-bit up counter and a three-bit up/down counter with various control lines are designed. Synchronous counters are used to synthesize precision waveforms for a variety of applications.

*Shift Register* A register is called a shift register when all bits stored in the register can move one position to the left or right of their present position by asserting control lines. A shift register holds the bit values until a shift command is received. This is why the register flip-flops are D flip-flops with the load function. Commonly used one bit shift operations are defined and implemented using the ASM method. Shift registers implement special state machines whose state numbers and outputs do not need decoding. This results in increased performance, because there are no decoding delays.

*Linear Feedback Shift Register* - **LFSR** A primitive polynomial represents an LFSR. Some applications of LFSRs are non sequential counters, pseudo random number generators, dividers of polynomials, and serial codeword encoders and decoders.

# 8 Memory

Everything useful is operated by a state machine, and every state machine uses memory. The memory may be just a one bit register, or $2^{25}$ bits supporting a central processing unit (cpu) that is one form of state machine. In any digital system a defined group of 1 to k bits represents a *word* of data stored in the memory. Memory stores words for use on demand.

A memory black box is connected to an address bus, an input/output (i/o) data bus, ASM datapath control lines, and power supply lines. This is the minimum set of connections (Figure 810 page 162). The memory may have more than one i/o port. The memory may consist of just one memory chip or be a complex assembly of memory chips, address decoders, buffers, and so forth. Nevertheless memory may be considered as a black box that may be analyzed as one equivalent memory chip. The address line bus presents an n-bit binary number to the black box. This number ultimately selects some k-bit stored word to be processed by commands on the control lines. After about half of the access time expires the addressed word is selected and available (inside the memory) for read or write operations that use up the second half of the access time. The word may have been stored at a prior time, and is now read from the memory and placed on the k bit output q data bus. Or the word is now a nothing, and is overwritten by a new word available on the k bit input data d bus.

***Access time*** The primary memory parameter is the *access time* that is the time required to execute a read or write operation. All memory chips allow access to any internal word register by giving each word a unique address. Access time is the time required to locate and read from, or write a word to, a memory. All read and write operations complete within the access time for all addresses. The access time is independent of the sequence of addresses, or the address of words being read or written. This is referred to as *random access capability*.

***Memory chips*** A memory chip is either a random access *read/write* RAM or a random access *read only* ROM. There are many forms of RAMs and ROMs. Words in a RAM are volatile. Turn the power off and then on. The array of cells now stores 1's and 0's in some unknown order so that the stored words are garbage. *The RAM cells always have a 1 or a 0 stored in them.* On the other hand the words in a ROM are permanent (nonvolatile). ROM words are written when the ROM is manufactured. EEPROMs are electrically reprogrammable ROMs.

*Static and dynamic* RAMs are either static RAMs (SRAMs) or dynamic RAMs (DRAMs). The one-bit cells in an SRAM are latches that are stable circuits that can hold data as long as the power supply is on. The one-bit cells in a DRAM are essentially a capacitor that loses its charge, representing a 0 or 1, over time.. That is why DRAMs are refreshed.

*Timing* The basic simplicity of memory chip timing is obscured by the understandable multitude of timing parameters found in data books. The timing parameters arise from propagation delays through the logic and *the timer in the memory chip* that controls read and write operations.

*Memory chip* A memory chip is basically an assembly of an address decoder, an array of one bit storage cells grouped as k bit words where k is a power of 2, a two way input/output circuit with k data lines known as d/q (in/out), and an internal timer *nobody ever mentions*. A realistic view of the memory chip is this. There is the address side of *n* address lines that select a word of k one bit cells from the array of one bit storage cells. There is the i/o side of k d/q (in/out) lines connected to the bit lines of the selected k one bit cells. A reasonable approximation is that the word addressing, and read or write i/o processes each use up one half the access time.

*Decoder* The n bit address drives a decoder that asserts one of $2^n$ word lines. The asserted word line has the same number as the address. Suppose address $1101\ 1010\ 0011\ 0111\ 1100_2$ is presented, and the access time is 6ns. Then 3ns later the word at this address is available for read or write operations.

*i/o read* On a read operation the 1 and 0 contents of the selected k one bit cells are placed on bit lines and subsequently delivered within 3ns to the d/q output terminals. Total access time is within 6ns.

*i/o write* On a write operation the 1's and 0's on the k d/q lines are stored in the selected word's k one bit cells via the bit lines within 3ns so that the total access time is again within 6ns.

*One bit cells* If the memory cell array has $2^{25}$ bits (33,554,432 bits), and the cells are organized as $2^5$ (32) bit words, then the memory can store data in $2^{20}$ words (1,048,576 words). This is referred to as a 32 mega*bit* memory, or a 4 mega*byte* memory, or a 1 mega*word* 32 bit word memory.

# 8.1 System Timing Waveforms

*A Central Processing Unit (cpu) Bus Protocol* provides a realistic context that shows how access time affects performance. Many computers use the following bus protocol where a read or write transfer cycle requires a minimum of three cpu bus clock periods. (There are many bus protocols.)

***Wait periods*** On read or write, the cpu tests for the presence of *brdy* at every rising clock edge (Figure 801). *Brdy is sent by the memory system.* If brdy is not detected then the clock period that just expired is referred to as a wait period or wait state. For example if the cpu bus clock period is 4ns and the memory access time is 14ns, then there are three wait clock periods (Figure 803). This is why memory chip access time is crucial.

***Read*** The cpu keeps control of the bus after the address, the *ads* start command, and the read signals are issued concurrently in bus clock period 1. Then the cpu waits until the memory system emits a brdy report. Brdy is the *data available* signal. While waiting, the cpu checks for brdy on every rising clock edge. On some rising clock edge the cpu detects brdy, reads in the data, and then completes the read (Figures 801, 802, 803). The state numbers $S_n$ are taken from the SRAM ASM chart (Figure 809 page 157).

**Figure 801 Cpu Read Cycle**

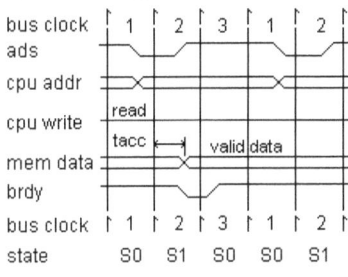

**Figure 802 Cpu Read Cycle, 1 wait state**

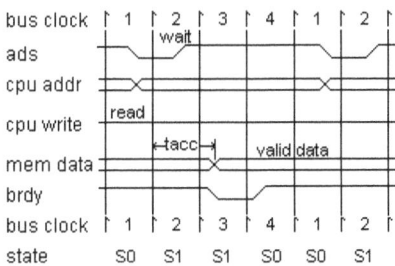

**Figure 803 Cpu Read Cycle, 3 wait states**

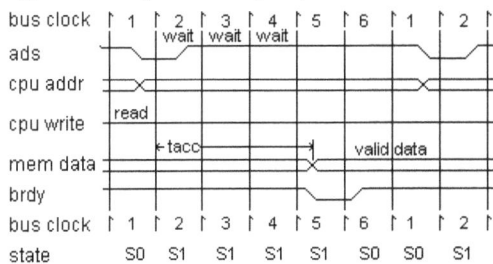

Digital Design

*Write* The cpu keeps control of the bus after the address, the *ads* start command, and the write signals are issued concurrently in bus clock period 1 (Figure 804). The cpu emits the data in bus clock period 2. The cpu waits until the memory sends a brdy report signaling that data has been stored. While waiting, the cpu checks for brdy on each rising clock edge. On some rising clock edge the cpu detects brdy and then completes the write (Figures 804, 805, 806). The state numbers $S_n$ are taken from the SRAM Read/Write ASM chart (Figure 809).

**Figure 804 Cpu Write Cycle**

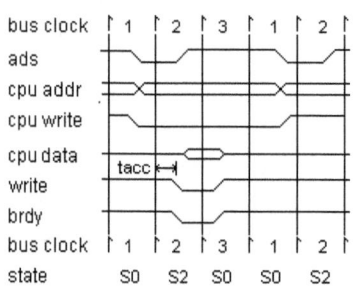

**Figure 805 Cpu Write Cycle, 1 wait state**

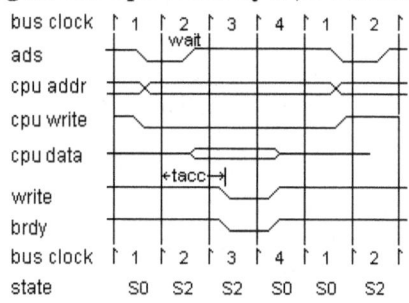

**Figure 806 Cpu Write Cycle, 3 wait states**

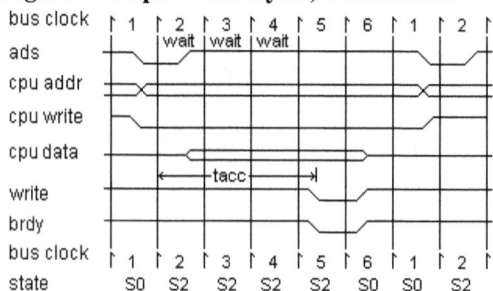

---

*Problem* 801 Reference Figure 806 where write cycle time is 6 clock periods, and there are three wait states. Redraw with two wait states.

---

**Figure 809 SRAM Read/Write ASM**

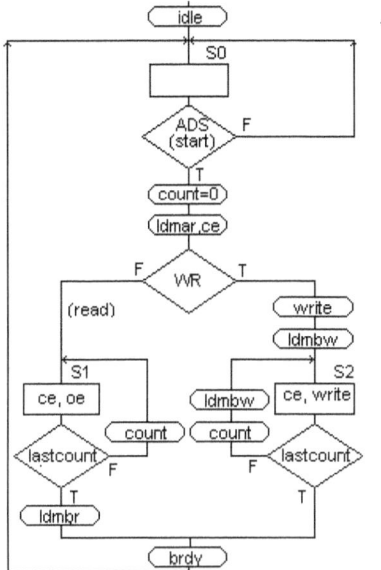

# 8.2 Basic System Design

We design the ASM charts for a memory system and a cache memory system. And, we explain how all the pieces fit in and how they function.

Every digital system includes a state machine controlling a *datapath*. The memory system in the datapath is defined by the data storage requirement. A basic memory system has an SRAM (static ram) or DRAM (dynamic ram) memory chip array of one or more chips, an address decoder to select the addressed word, and registers to store address and data.

*Cache is one level of complexity* Typical low cost DRAM access time is say 30ns, while SRAM access time is say 8ns. A SRAM *cache* memory is backed up by a memory consisting of DRAM chips. The combination has an average access time of say 12ns. Two levels of cache are common, and three levels are not unheard of (Section 8.3 page 165).

*ECC is another level of complexity* A high reliability system includes error correction and control (ecc) circuits (Section 8.4 page 176).

## 8.2.1 Address Decoder

***One chip memory*** A memory may consist of just one chip. In that case a system address decoder is not required, because every memory chip has an internal decoder that can access every word. Connect the system address lines directly to the chip address line inputs to access any word in the chip.

***Word size*** Memory chips are organized to store words that are assembled from 1, 2, 4, 8, 16, 32, 64 or more bits. If a 44 bit word *memory system* is assembled with 8 bit and 4 bit word chips, then 6 memory chips (5×8+1×4=44) are *equivalent to one* 44 bit word memory chip (Figure 807). The memory design proceeds on the basis of this equivalent 44 bit word chip. In other words the decoder is not affected by individual chip word size (In Section 8.4 page 176 we explain why word size might be 44 bits when the data word is 32 bits).

**Figure 807 44 bit chip equivalent**

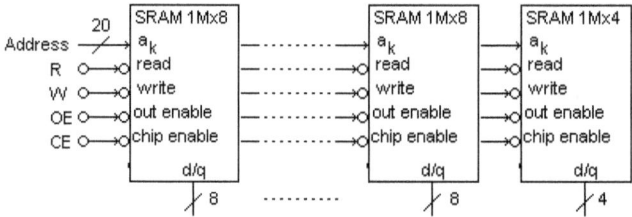

***Decoder*** A large memory system *address space* addresses a group of address *blocks*. Each block addresses a group of address *banks*. Each bank addresses a group of memory chips. *The address space hierarchy is: blocks - banks - chips.*

We start the memory decoder design by selecting a memory size, and take it from there. An 8 megabit memory contains $2^{23}$ bits (8,388,608 bits).

We find that hexadecimal (hex) numbers are easier to work with when the numbers are as large as $2^{23}=1000\ 0000\ 0000\ 0000\ 0000\ 0000_2=800000_{16}$. This means 23 address lines are required to address the 8 megabit memory. Note: the number of zeros after the 1 equals the exponent. Try $2^5=100000_2$.

Assume an SRAM contains $2^{20}$ 44 bit words ($2^{20}=1,048,576$). The hex number $100000_{16} = 2^{20}$, where $100000_{16}$ represents the one megaword address range from $00000_{16}$ to $FFFFF_{16}$. Add 1 to FFFFF to get $100000_{16}$.

The 1 megaword SRAMS address range spans a sequence of numbers representing the *addresses* $00000_{16}$ to $FFFFF_{16}$. In other words our *bank size* is $2^{20}=100000_{16}$ words.

A bank address has 20 address lines, $A_{13\ hex}$ to $A_{00\ hex}$, corresponding to the 20 bits in the addresses. Three lines, $A_{16\ hex}$ to $A_{14\ hex}$, address the 8 banks. If there are 8 blocks, three more lines, $A_{19\ hex}$ to $A_{17\ hex}$, are required. For example address lines $A_{19}$ to $A_{14}$ ($011101_2 = 1D_{16}$) select block 3 $011_2$, bank 5 $101_2$ (Figure 808).

**Figure 808 Decoder for 8 megaword address space**

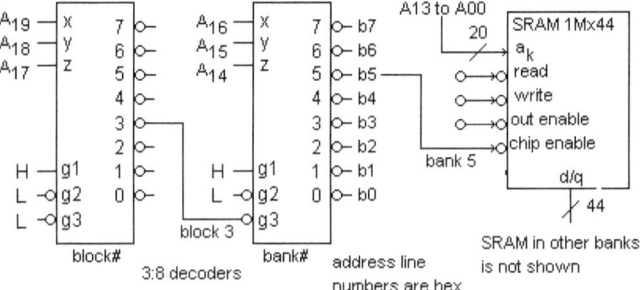

When analyzing the 8 megaword decoder (Figure 808) we take careful note of which numbers are base 2 and which are base 16. As we count hexadecimal numbers up from 0, every time we recycle from $FFFFF_{16}$ to $00000_{16}$ we increment bank bits $A_{16}A_{15}A_{14}$. While in bank 5, $A_{16}A_{15}A_{14} = 101_2$ and the 3:8 bank decoder output $b_5$ is asserted (Figure 808).

In the same way, while in block 3, $A_{19}A_{18}A_{17} = 011_2$ the 3:8 block decoder output 3 is asserted. Therefore bank 5 of block 3 is active when the most significant address bits are 011 101 xxxx xxxx xxxx xxxx $xxxx_2$.or 01 1101 xxxx xxxx xxxx xxxx $xxxx_2$. I.e. the address range for bank 5, block 3 is $1D00000_{16}$ to $1DFFFFF_{16}$. *Clearly great care is required when dealing with addresses such as these.*

***Bits, bytes, and words*** There is potential for confusion, because memory chip size is specified as so many bits, or as so many bytes (of 8 bits) , or as so many words (of 32 bits). We prefer bits, such as 4 megabits ($2^{22}$).

Digital Design

## 8.2.2 Registers

The address and the word to be written must be stored in registers, while access time elapses. The word read from the memory is stored in a register after access time elapses.

The generic names for these registers are *memory address register* (MAR), *memory buffer write* (MBW), and *memory buffer read* (MBR) (Figure 810 page 162).

The commands *ldmar, ldmbw, ldmbr* store words in the MAR, MBW, and MBR. In a computer system, the MAR. MBW, and MBR are part of the cpu chip. The MBR and MBW are merged in many computer chips.

---

*Problem* 802 Reference Figure 808. Show that the address range for bank 5, block 3 is indeed $01D0\ 0000_{16}$ to $01DF\ FFFF_{16}$.

*Problem* 803 Address lines are $A_F$ to $A_0$. Design a decoder spanning the address space with 8K ($2^{13}$) banks. Hint make a truth table.

*Problem* 804 Address lines are $A_F$ to $A_0$. Design a read/write decoder for i/o ports at addresses 7BE0, 7BE1, 7BE2, 7BE3. Hints that addresses lines are constant? that vary?.

*Problem* 805 Address lines (hex) are $A_{17}$ to $A_{00}$. Design a decoder for hex bank address spaces A00000 to A0FFFF, B00000 to B0FFFF, C80000 to C8FFFF, D80000 to D8FFFF. What is the RAM size? Why?

*Problem* 806 Address lines (hex) are $A_{1F}$ to $A_{00}$. Design a decoder for hex bank address spaces 3BA00000 to 3BA3FFFF, 3BB00000 to 3BB3FFFF, 3BC80000 to 3BCBFFFF, 3BD80000 to 3BDBFFFF. What is the RAM size? Why?

*Problem* 807 Address lines (hex) are $A_{1F}$ to $A_{00}$. Design a decoder for hex bank address spaces 590B0000 to 590CFFFF, 598B0000 to 598CFFFF. What is the RAM size? Why?

---

### 8.3.3 ASM Design for a Memory System

ASM design requires learning about memory chip specifications, and cpu read and write timing diagrams. This leads into how to design the memory datapath. Memory data path control inputs are ASM outputs, and data path outputs are ASM inputs (Figures 809, 810 page 162). The memory stands by in state 0 waiting for a cpu *ads* command. Read and write states control the read and write processes.

The start of a read or write cycle is signaled by the *ads* signal. In the same clock period memory control also produces the *address* and *read or write* signals (Figures 802, 805 where *cpu write* is H on read and L on write). In a write cycle the *data* appears on the data bus in the next clock period 2 (Figure 805).

Data read from memory appears when the *access time* expires, and in the same clock period the memory system *brdy* signal signals that data is available (Figure 802). Memory chips do not have an *access time* timer. This why an access time timer is added to the memory system. For example, the counter in Figure 810 measures the access time by issuing *lastcount* that asserts brdy.

In a write cycle the source's data has to be available in the same clock period in which the *access time* expires (Figure 805). The *write* signal enables data to be written into the memory on the next clock edge. In the same clock period the memory system *brdy* signal signals that data has been written. Brdy marks the end of a read or write cycle.

**Figure 802 Cpu Single Transfer Read Cycle, 1 wait state**

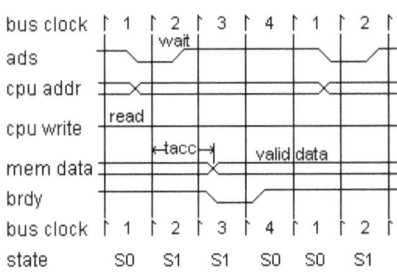

**Figure 805 Cpu Single Transfer Write Cycle, 1 wait state**

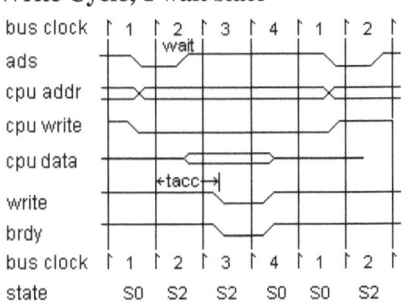

***Read Cycle*** The memory system idles in state $S_0$ when input ADS is false. At the start of a read cycle ADS is asserted, and after a gate delay conditional outputs *count=0*, *ldmar*, and *ce* (chip enable) are asserted (Figures 809, 810, 811). The ASM moves to state $S_1$ on the next clock edge, because the exit path from $S_0$ is via diamond exits ADS=T and WR=F. In $S_1$ unconditional outputs *ce* and *oe* (output enable) are asserted. Observe the continuous waveform for *ce* asserted in $S_0$ and $S_1$ (Figure 811). Memory chips require *oe* to be asserted on read. Counter input *count* is asserted while in $S_1$ AND *lastcount* is F. Back in $S_0$, the counter was set to 0 by *count=0*. Therefore *count* advances the count from counter state 0 to 1 to 2, when *lastcount* is asserted by design. After a gate delay *lastcount* goes to T, and *ldmbr* and *brdy* are asserted.. The exit from $S_1$ to $S_0$ occurs on the first clock edge after *lastcount* is asserted. That ends the read cycle.

**Figure 809 SRAM Read/Write ASM**     **Figure 810 Memory system**

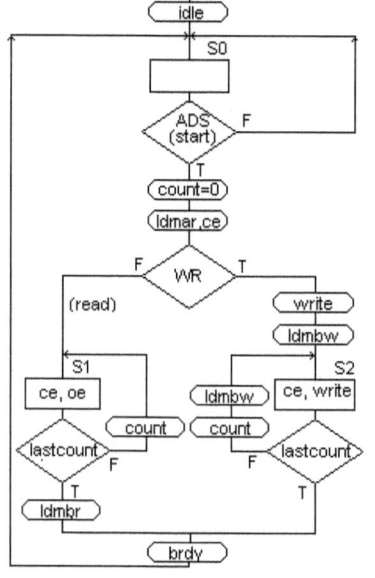
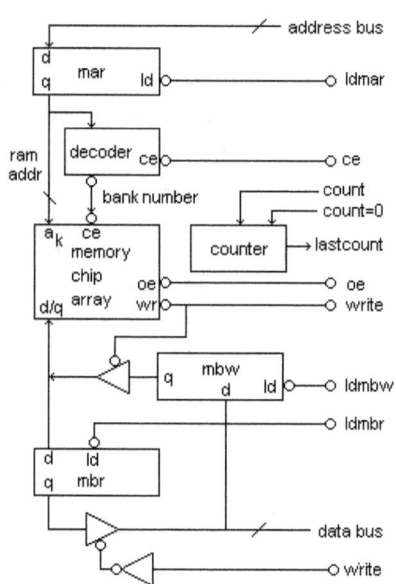

*Write Cycle* At the start of a write cycle ADS is asserted, and after a gate delay conditional outputs *count=0*, *ldmar*, and *ce* are asserted (Figures 809, 810, 811). In the same $S_0$ clock period outputs *write* and *ldmbw* are asserted, because WR=T. The ASM moves to state $S_2$ on the next clock edge, because the exit path from $S_0$ is via diamond exits ADS=T and WR=T. In $S_2$ unconditional outputs *ce* and *write* are asserted. Observe the continuous waveforms for *ce, write, ldmbw* asserted in $S_0$ and $S_2$ (Figure 811). Counter input *count* is asserted while in $S_2$ and *lastcount* is F. Back in $S_0$, the counter was set to 0 by *count=0*. Therefore in $S_2$ *count* advances the count from counter state 0 to 1 to 2, when *lastcount* is asserted by design. After a gate delay *lastcount* goes to T and *brdy* is asserted.. The exit from $S_2$ to $S_0$ occurs on the first clock edge after *lastcount* is asserted that ends the write cycle.

**Figure 811 Memory System Timing Diagram – read followed by write**

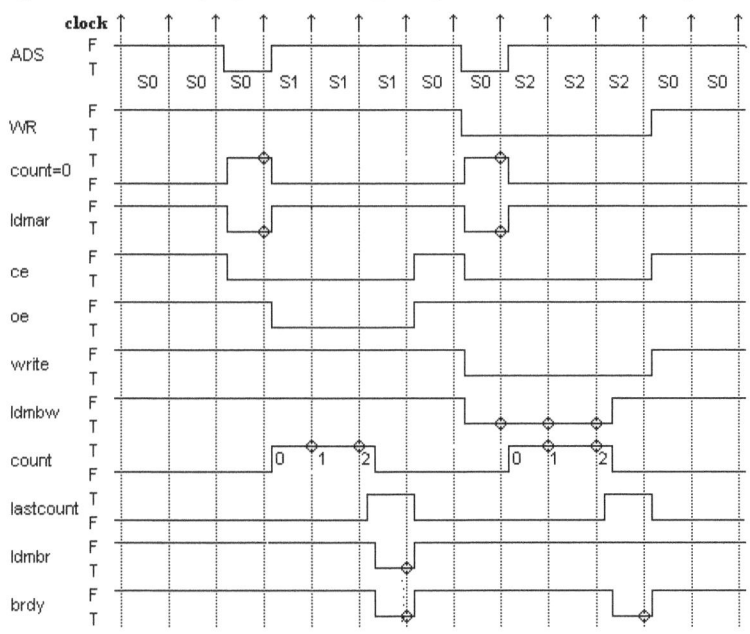

The circles show when a rising clock edge executes the command.

---

*Problem* 808 Reference Figure 809. Draw a finite state machine black box representing the ASM. Show the inputs and outputs

*Problem* 809 Reference Figure 809 and problem 808. Design the finite state machine.

---

## The ASM Chart and the Finite State Machine

The ASM chart represents a synchronous finite state machine (FSM). The ASM chart is drawn using lines, the state rectangle, the input diamond, and the conditional output oval (Chapter 5).

• *States* are represented by rectangles.

• *Inputs* are inside diamonds that have T (true) and F (False) exits.

• *Clock* is hidden. The clock action is as follows. At every rising clock edge the next state decision is to move from the present state rectangle to the next state rectangle. The present state and the inputs' values (T or F) at the clock edges determine which state is the next state. In Figure 809 page 162, when the present state is $S_1$, if *lastcount* is F the next state is $S_1$. Just follow the path from $S_1$'s exit through the *lastcount* diamond back to $S_1$'s entrance. Ovals, such as the one with the *count* command, are ignored when tracing paths to the next state. In this example when *lastcount* is F(alse) the *count* command executes on the rising clock edge that also steps to the next state $S_1$. Note that the present state can be the next state (in this example $S_1$ steps to $S_1$ on every, repeat every, rising clock edge as long as *lastcount* is F).

• *Unconditional outputs* are listed in state rectangles. Unconditional outputs are asserted for every clock period the associated state is the present state. In Figure 809, while the present state is $S_1$ unconditional outputs *ce* and *oe* are asserted.

• *Conditional outputs* are shown in ovals. Conditional outputs are asserted for every cycle the path from the present state to the oval entrance is asserted. In Figure 809, the *count*, and *ldmbw* conditional outputs are asserted when the present state is $S_2$, and *lastcount* is F.

*Timing diagrams* A timing diagram (e.g. Figure 811 page 163) shows, in graphical format, circuit logic levels as a function of time. The horizontal axis represents time, and the vertical axis represents logic levels high and low (H and L). Each trace on a timing diagram shows when one logic variable changes level from true T to false F, and vice-versa. For example, memory timing diagrams show the signals on the cpu external bus as functions of time, and time is measured in bus clock periods.

# 8.3 Memory Hierarchy Design

Cost of memory chips is a, if not the, major reason for using memory hierarchies. A memory hierarchy organization is based on the fact memory chip cost per bit is inversely proportional to bit access time. Cost per bit increases dramatically as access time decreases. The question before us is how to do we set up a hierarchy?

Computers and their memories exist to execute programs. So how do programs execute? Program execution focus is on some part of the address space in any short time interval. The address space contains instructions and data, and the focus is local in time and, it turns out, also local in address. If an instruction or data is referred to, then most likely that instruction or data will be referred to sooner than later. If an instruction or data is referred to, then most likely other instructions or data at nearby addresses will be referred to sooner than later.

In other words program execution focus is on a *portion* of the address space in any short interval of time. Portion infers a smaller memory will do as long as access to the entire memory is available. Industry decided that smaller made short access time chips affordable. Consequently a smaller, higher cost, faster SRAM memory working in conjunction with a large, lower cost, slower DRAM memory became the basic memory hierarchy. And so memory systems were designed to include memories with various access times interacting with each other.

A memory hierarchy consists of a series of memories ranked according to access time. $Cache_1$ has the shortest access time. (Note: *level* is used as an alternative to *cache*.) When a cpu with a memory hierarchy issues a memory access request, the request goes to $cache_1$. If the addressed word is found in $cache_1$, then there is a *hit*. The read or write operation is executed in $cache_1$, and the cpu moves on. If the addressed word is not found in $cache_1$, then there is a *miss*. On a miss the addressed word is looked for in $cache_2$. If found, then there is a hit. The read or write operation is executed in $cache_2$. Then some block of words stored in addresses including the address just used is transferred from $cache_2$ to $cache_1$ in anticipation of being used soon. If the addressed word is not found in $cache_2$, then there is a *miss*. On a miss the addressed word is looked for in $cache_3$, and so forth until the bottom cache that is the main memory, is reached.

## 8.3.1 Cache Data Path Architecture

Cost reduction requires that cache size is less than main memory size. If cache$_1$ size is 512K ($2^{19}$) words of 32 bits, and the main memory size, is 1024M ($2^{30}$) words, then there are $2^{30-19} = 2^{11}$ or 2048 *banks* of 512K words in the main memory. Assume the cpu has 32 address lines $A_{31\_10}$ to $A_{0\_10}$.

Define a tag address as $A[31_{10}:19_{10}]$, and an *index address* as $A[18_{10}:0_{10}]$. A cache stores tags, and cpu data words (Figure 812).

$$| tag \qquad\qquad | index$$
$$| A_{31}A_{30} .... A_{19} \quad | A_{18} .... A_0$$

*The tag is the bank address of the word. The index is the word's address within the bank. The concatenated tag address and index address is the word's main memory address.*

The Tag SRAM stores tag addresses $A[31_{10}:19_{10}]$.
The Word SRAM stores the tagged cpu data words.

**Figure 812 Simplified Cache$_1$ Memory Diagram**

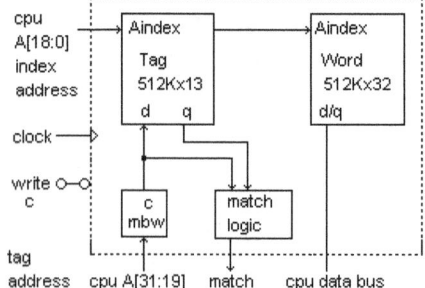

When a cpu writes a word into the *Cache Word SRAM,* it also writes the word's tag address into the *Cache Tag SRAM.* The tag and the word are stored in their rams at the index address $A[18_{10}:0]$. Then when a cpu issues a read address the cache system reads from the index $A[18:0]$ address the tag address $A[31:19]$ and the associated word. This is automatic, because the cpu address lines have been *wired* to as shown in Figure 812.

If the stored tag equals the tag in the cpu address, then the match report is a 1 and we have a hit. The word is taken in from the cpu data bus by the cpu, because a hit means the word the cpu is addressing is in the cache.

## 8.3.2 Cache Control

Assume that there is only one cache, $cache_1$, in the computer system in addition to the *DRAM main memory*. The computer's operating system program loading process copies a program from the hard drive to the DRAM memory. Then the operating system forces a jump to the start of the program placed in the DRAM memory. As the program executes $cache_1$ fills, because initially every read is a miss. Ultimately the running program always finds $cache_1$ filled with copies of words from various addresses in the DRAM memory. Here is how $cache_1$ fills with data, and subsequently changes stored data. (The ASM in Figure 816 page 171 implements the read and write policies.)

***Read Policy*** On a hit, data is read from $cache_1$ and delivered to the cpu. On a miss, data is read from the DRAM memory, and written to the $cache_1$ Word SRAM as it is delivered to the cpu. At the same time the associated tag is written to the $cache_1$ Tag SRAM.

***Read Action*** Program code includes instructions to fetch data or other instructions from memory. *The cpu emits a read command to all memory levels in parallel.* $Cache_1$ is checked for presence of the desired word. The check is made by reading $cache_1$ to get a match report. The tag read from the tag ram (Figure 812) is compared to the tag in the cpu address and produces a match report. If the desired word is in $cache_1$ then the match report is true. This is a read *hit*. If its a hit then the DRAM memory read command is aborted, and the copy of the word in the $cache_1$ word SRAM is delivered to the cpu. This ends the read cycle.

If the desired word is not in the cache the match report is false and its a read *miss*, then the DRAM memory read cycle proceeds to completion to deliver the new word to the cpu, to write a copy of the new word, and the new word's tag to $cache_1$.

***Write Policy*** Since the cache is always full, writing to the cache replaces some other word. There are various strategies for selecting the word to be replaced. The fastest strategy is to replace the word at the current index address.

On a write hit words are stored in the SRAM *and* the DRAM. On a write miss words are stored in the DRAM, and the cache is not modified, because a write miss means there is no stale data in $cache_1$ to be replaced.

Digital Design

*Write Action* All cpu write commands are sent to the DRAM memory and are executed to completion. In parallel, cache$_1$ is checked for a match, while the (slower) DRAM memory is processing a write command. If match is true (its a write hit), then the new data word is also written to cache$_1$ thereby replacing the now obsolete data word. If its a miss cache$_1$ is not changed.

> Emphasis: *a hit reports a tag address match, not a data match.*

*Cache controller* The cache controller is in charge of the memory system. This relieves the cpu. The cache controller responds to a cpu read or write command, and returns *brdy* (and a word on reads) when its action is completed. The cache controller is the black box that interacts with the cpu. A table shows the required actions based upon hit and miss results. A simplified block diagram shows how the cpu, the cache controller, the SRAM cache memory and the DRAM memory interact (Figure 813).

| result | sram | dram |
|--------|------|------|
| *read hit* | *read* | *no action* |
| *read miss* | *write* | *read* |
| *write hit* | *write* | *write* |
| *write miss* | *no action* | *write* |

**Figure 813 Simplified CPU Memory System Diagram**

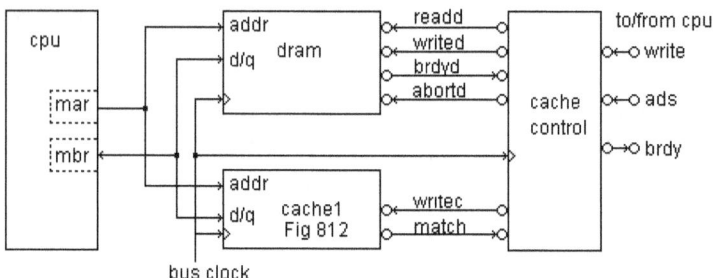

Every cpu has an MAR (memory address register) and an MBR (memory buffer register) for data i/o. The MAR holds the current word address. The MBR sends and receives data on writes and reads respectively. Tri-state gates preclude bus contention. On reads the cpu simultaneously sends a read command and the corresponding address. On writes the cpu simultaneously sends a write command, a word, and the address.

168

### 8.3.3 Read Hit and Miss Cycles

A memory read cycle transfers one word from memory to the cpu. The cpu begins a read cycle when it asserts *ads* (the address strobe). Concurrent with *ads*, the cpu outputs the address. Then the cpu tests for *brdy* at each rising clock edge. System logic issues the brdy report when the data is available so that the cpu knows when to complete the transfer. What happens in the interim depends on whether a hit or miss occurs, and also depends on the memory SRAM and DRAM access times. Observe the sequence of states and the count in the timing diagram (Figure 814 page 170). System logic must wait for SRAM access time *tacc* to expire before the first word is available (SRAM *tacc* is elapsed time from clock edge to data available). Here is a detailed discussion of a read transfer (Figures 814, 816 pages 170, 171).

***Bus clock period* 1** This is the start period. Assume the ASM is idling in $S_0$, and *refresh=F* (Figure 816). Conditional output *count=0* is asserted. When *ads=1* AND *busy=0* AND *write=F*, the conditional output *read d* (read DRAM) is immediately asserted. On the next bus clock edge the *read d* command starts a DRAM read cycle controlled by a *linked* ASM (Figure 817 page 173). On the same clock edge the ASM steps from state $S_0$ to state $S_1$ where output *read c(ache)* starts a cache SRAM read cycle.

Emphasis: *Count=0* resets the system logic access time timer to zero. *Read d* starts a DRAM read cycle just in case this read is not a cache hit.

***Bus clock periods* 2, 3, 4, 5** The reads continue (Figures 814, 816). While the ASM dwells in $S_1$, the SRAM and DRAM are accessing the data, and *count* is incremented on each clock edge. The *c done* report goes true in period 5 when SRAM access time *tacc* expires (*tacc* is defined by count=3, Figure 814), and valid *data* becomes available. By this time *match=T* if this is a hit. That means a DRAM read is not required. Therefore *abort d* is asserted to abort the DRAM read, and *brdy* is asserted. If match is F this is a miss, and the ASM steps to state $S_3$ where the DRAM continues reading. When DRAM data becomes available (*d done=T*) a *write c* command writes the word and tag into the cache, and *brdy* is asserted.

*Bus clock period* **6** After three wait periods and brdy (Figure 814), period 6 is the last bus clock period of the cpu's read cycle on a hit. The memory system resumes idling in state $S_0$ waiting for the next *ads* signal.

The memory system resumes idling in state $S_0$, because *busy=1* (Figures 815, 817 page 173). The system is waiting for *busy=0* from the DRAM ASM (Figure 817) *reporting the end of the DRAM cycle.*

**Figure 813 Simplified Memory System Diagram**

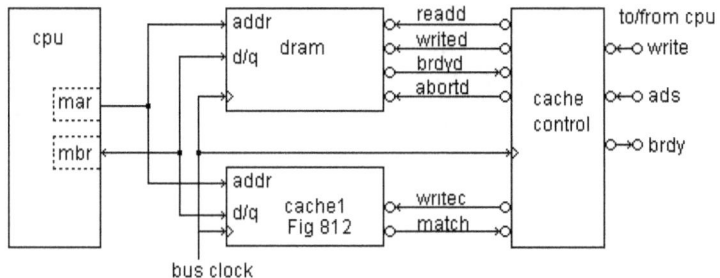

**Figure 814 Single Transfer Read Hit Cycle, 3 wait states**

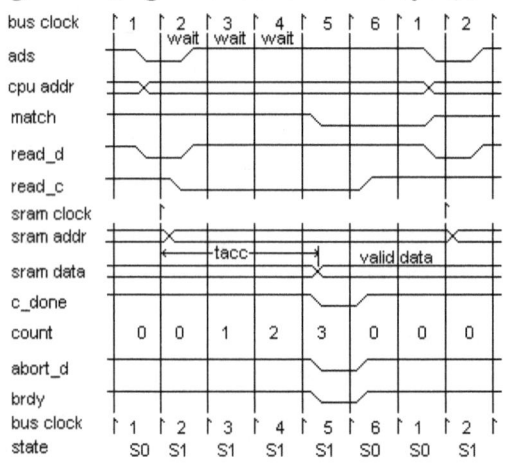

---

*Problem* 810 Reference Figure 816. Write a cycle by cycle description of a read miss cycle. Make a timing diagram similar to Figure 811 page 163.

---

## Figure 815 Single Transfer Write Hit Cycle, 3 wait states

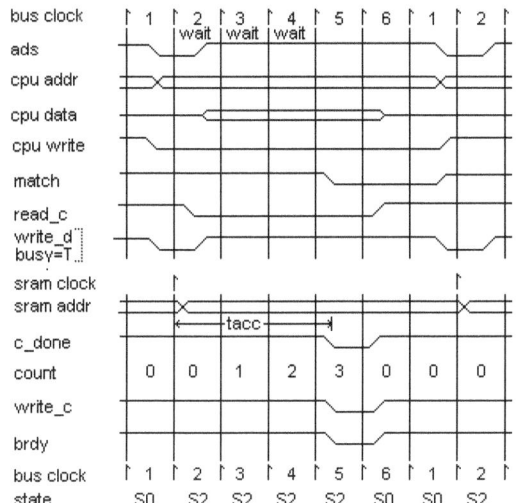

## Figure 816 ASM Chart for Memory System Algorithm

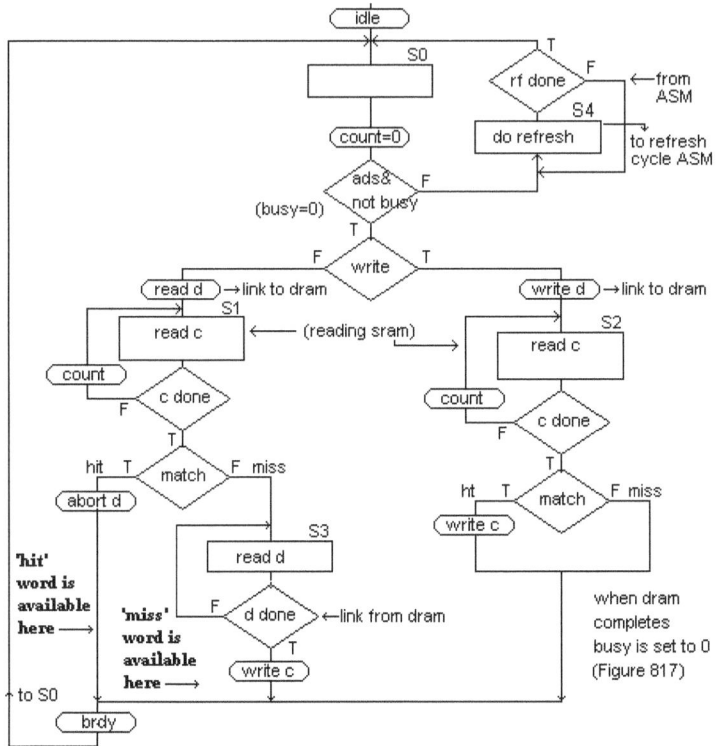

## 8.3.4 Write Hit and Miss Cycles

A memory write cycle transfers one word from the cpu to the memory. The cpu begins a write cycle when it asserts *ads* in bus clock period 1. Concurrent with *ads*, the cpu outputs the cpu address, the cpu write signal and, in clock period 2, the cpu data word to be stored (Figure 815 page 171). Then the cpu tests for *brdy* at each rising clock edge. System logic issues the *brdy* report when the data is stored so that the cpu knows when to complete the transfer. What happens in the interim depends on whether a hit or miss occurs, and also depends on the SRAM and DRAM access times.

Observe the sequence of states, access time *tacc*, and the count in Figure 815. System logic must wait for SRAM access time *tacc* to expire before writing. Here is a detailed discussion of a write transfer (Figures 815, 816 page 171).

***Bus clock period* 1** This is the start period. The ASM is in $S_0$ (Figure 816), *refresh=F*, and conditional output *count=0* is asserted. *Count=0* resets the system access time timer to zero. When *ads=1* AND *busy=0* AND *write=T*, the conditional output *write d* is immediately asserted. *Write d* starts a DRAM write cycle regardless of a hit or miss in order to always have correct words in the main memory. On the next clock edge the ASM steps from state $S_0$ to state $S_2$.

***Bus clock periods* 2, 3, 4, 5** While the ASM dwells in $S_2$, the cache SRAM is accessing the data in it, and *count* is incremented on each clock edge. The *c done* report goes true in period 5 (Figure 815) when the access time *tacc* expires (count equals 3). By this time *match=T* if this is a hit. A hit means the data is now stale, and must be replaced by the new cpu data that is in the process of being written into the DRAM. *Write c* writes the new word into the cache's word SRAM. If this is a miss *write c* does not occur. At this point the cache transfer is complete and *brdy* is asserted.

***Bus clock period* 6** After three wait periods and brdy (Figure 815), period 6 is the last bus clock period of the cpu's write cycle. The memory system resumes idling in state $S_0$, because *busy=1* (Figures 816, 817). It is waiting for *busy=0* (Figure 816) *reporting the end of the DRAM write*.

---

*Problem* 811 Reference Figure 816. Write a cycle by cycle description of a write miss cycle. Make a timing diagram similar to Figure 811 page 163.

---

### 8.3.5 The ASM Chart for DRAM Read/write Cycles

The expression for the *do* input (Figure 817) is *do=read d OR write d*. On *do=T* the conditional output *busy=1* sets the busy flip-flop to state 1, the ASM steps from $S_{100}$ to $S_{101}$ where the DRAM starts a read or write cycle.

The *write d* signal from the main ASM also sets the *write* flip-flop so that *write d* is available at state S104. The *write d* flip-flop is reset in state $S_{107}$ (Figure 817).

State $S_{107}$ outputs *busy=0* to reset the *busy* flip-flop, and *d done=1* goes to the main ASM (Figure 816) to end the read cycle, which steps to $S_0$.

On a DRAM read or write the main ASM waits for the *busy=0* report from the link ASM $S_{107}$ that allows the next cycle to start (Figure 816).

***Busy flag*** The *busy* flip-flop is set to 1 by the linked ASM (Figure 817) when a DRAM read or write cycle starts.

*Busy=1* means *the cache or DRAM is still writing or reading* as a result of a *prior write or read command, and the system waits.* (Figure 816 page 171). The system waits in $S_0$ on writes and in $S_3$ on reads because *busy* is still 1 or *d done=1* has not been issued. When the prior DRAM memory cycle ends busy is set to 0 and d done to 1 by the linked ASM state $S_{107}$ (Figure 817). Then the next read or write cycle can execute.

Busy is not *brdy*.

**Abort d** We do not show the *abort d* diamond at the output of all states (Figure 817). The abort d input resets the link ASM to state $S_{100}$, and sets busy and write to 0.

**Figure 817 ASM Chart for DRAM read/write cycles**

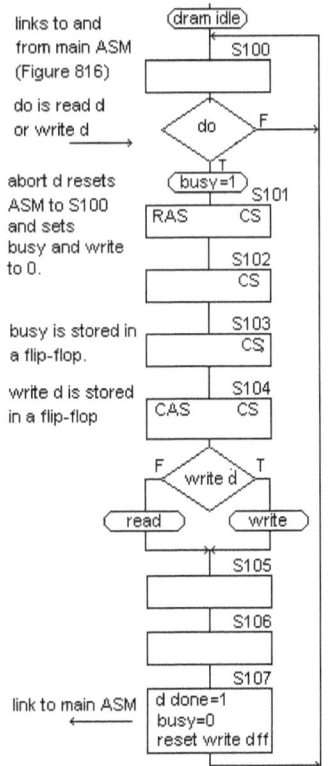

links to and
from main ASM
(Figure 816)

do is read d
or write d

abort d resets
ASM to S100
and sets
busy and write
to 0.

busy is stored in
a flip-flop.

write d is stored
in a flip-flop

link to main ASM

## 8.3.6 DRAM Refresh Cycles

DRAMs are refreshed within some time $T_R$ so that data is not lost. Refresh is a third type of DRAM memory cycle.

The cells in one row of the DRAM row/column cell structure are refreshed in each refresh cycle. All rows have to be refreshed once per time interval $T_R$. If there are 4096 rows, then the 4096 rows are refreshed within time interval $T_R$. There are no rules specifying *when* a specific row is refreshed in any $T_R$ time interval.

Assume $T_R=10$ms$=10000\mu$s. If DRAM refresh cycle time $t_C$ is 45ns and there are 4096 rows in the one bit cell array then refresh cycles use up 4096 $t_C$ or 184,320ns or about 185$\mu$s. The 185$\mu$s is about 2% of $T_R$ that leaves 98% of the time for read and write cycles. Then if an independent timer in the memory control logic produces 4096 *refresh* signals every 10000$\mu$s all cells will be refreshed in time.

As long as *ads* $=F$ state $S_4$ will execute refresh cycles via a linked ASM that returns the *rf done* report (Figure 816 page 171). The refresh cycle circuit knows how much $T_R$ time remains and how many rows have been refreshed. If necessary it withholds the *rf done* report until all rows are refreshed.

The ASM structure (Figure 816) guarantees that the refresh process does not interfere with read and write cycles that are executed. The cpu may have to wait, while a refresh is underway. This is known as *interleaved refresh*. (We do not show the complex refresh logic circuit.)

Another form of refresh is *burst or self refresh* that simply refreshes all rows without pausing for read or write cycles. Self refresh makes the memory system unavailable for say 4096 $t_C$ or 184,320 ns. This time out is not acceptable in many systems.

---

*Problem* 812 Reference Figure 816 state $S_4$. Create an ASM chart for the refresh function.

*Problem* 813 Reference Figure 816 state $S_4$. Design the circuit for the refresh function.

---

## Recapitulation

**Read Algorithm** If the cache controller receives a read command (write=F) when the *busy flag* is 0 the controller starts a DRAM memory *read d* and a cache$_1$ *read c* (Figures 813, 814, 816 pages 170, 171). The cache$_1$ read completes, while the DRAM read continues in the linked ASM (Figure 817 page 173).

**Read Hit** If *match* reports a hit (Figure 816) during a read, then the DRAM read is aborted by *abort d* which resets the linked ASM to $S_{100}$ (Figure 817), resets *busy* and *write* flip-flips to 0, and a *brdy* report is forwarded by the cache controller to the cpu. The cpu receives *brdy* that tells it to take into its MBR the word just read from cache$_1$. This completes the cpu read cycle.

**Read Miss** If match reports a miss the DRAM read proceeds to completion setting *busy=0, d done=1* (Figure 817), then the cache controller issues a *brdy* report to the cpu (Figure 816). The cpu receives *brdy* that tells it to take into its MBR the word just read from DRAM. At the same time the cache controller issues a *write c* to cache$_1$ to store the word just read from the DRAM memory, and the tag at the same index address (Figure 812 page 166). This completes the cpu read cycle as the cache$_1$ write continues.

**Write Algorithm** The *write though* strategy is used here. If the *busy flag* is 0, then when the cache controller receives a write command it starts a DRAM memory write setting *busy* and *write* flip-flips to 1, and also starts a cache read (Figures 813, 815, 816 pages 170, 171).

**Write Hit** When match reports a hit a cache$_1$ *write c* is started to replace the stale word in the cache (Figure 816). The cache$_1$ write completes well before the DRAM write completes. So the cache controller forwards *brdy* to the cpu that completes the cpu write cycle (this is OK, because busy=1). The slower DRAM is still writing and when done sets the *busy* flag to 0.

**Write Miss** When match reports a miss the cache write is not started. Since the cache is done the cache controller issues a *brdy* report to the cpu that completes the cpu write cycle. However the slower DRAM comtinues its write cycle to completion. The *busy* flag was set to 1 to acknowledge this.

# 8.4 Error Correction and Control

The purpose of an error detection and error correction system is to detect and correct bits inverted from 1 to 0, and, or 0 to 1. If we know the positions of incorrect bits in a word, then we can correct them because we know which bus wires carry the bits in error. We know, because bits are numbered from 1 to n according to their position in an n bit code word.

**Figure 818 Encoding a word**

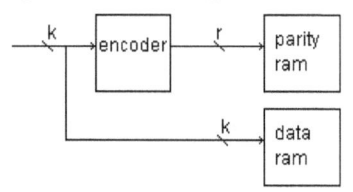

*Parity* If a k bit word is a message, then there are $2^k$ distinct messages (if k=4 there are $2^4$=16 possible messages). If any of the k bits is changed it becomes an error bit. The result is *another* k bit message. However this *received* message is an error, because it is not the message we sent. We cannot detect nor correct this type of error. This problem is avoided by *adding r parity bits* to the k message bits to create a *code word*. We can detect and correct code word errors, because *t or fewer errors* do not convert the code word into some other valid message. *However, more than t errors in a codeword can produce a codeword with one of the valid messages.*

***Linear binary block codes*** These are one of the most widely used codes. The cpu emits a block (word) of k binary digits in parallel. The encoding process *takes in a block* of k bits at a time. Then according to a set of r parity equations built into an encoder chip, the encoder calculates and appends r parity bits to the block of k bits to form a code word of n=k+r bits. The encoder can produce $2^k$ different n-bit code words with valid messages. The k bit word is stored in the data ram and the r parity bits are stored in the parity ram (Figure 818).

Errors can create $2^n-2^k$ code words that do not contain valid messages. Valid words stored in memory can be corrupted (bits are inverted 1 to 0, 0 to 1) for several reasons. Inverted bits are detected on reads, and corrected by a decoder (Figure 819 page 178).

The major binary block code parameters are n, k, r, and t, where t is the number of error bits the code can correct. The symbol for a block code is (n, k).

***R. W. Hamming's ingenious idea*** Parity equations are the focus of code design. R. W. Hamming's ingenious invention of codes[1] showed that parity bits can be used to detect error bits if parity bit values are derived from values of their associated message bits.

*Apply r parity check calculations to the received word R to produce r syndrome bits. Interpret these r syndrome bits as a binary number equal to the number of a position in the received word R thereby identifying the bit which is in error.*

There are $n=k+r$ bit positions in a Hamming code word: r parity bits and k message bits. Hamming restricted n to equal $2^r-1$ (we will see why in a moment). S is the name of the (decoding) syndrome matrix that has r rows. By design each row of S represents one of the r parity equations. The syndrome calculation produces r bits that represent one of $2^r$ binary numbers. If number zero is defined as zero errors, then in Hamming's design the numbers 1 to $n=2^r-1$ represent the n bit positions of the received word. This is why he restricted n to equal $2^r-1$. If $r=3$. then $n=2^3-1=7$, and $k=n-r=4$. The Hamming codes are limited to $t=1$ (can correct only one error). Another significant problem is that the $n=2^r-1$ equation is incompatible with $2^k$ word lengths such as 16, 32, and 64 (see BCH and the Example in upcoming paragraphs).

***H matrix*** The columns of H are binary numbers 001 through 111, because Hamming wanted the syndrome calculation to produce a bit position number. He showed that this happens when the parity equations are defined as follows. R is the code word as received.

(1) $H = \begin{bmatrix} 1 & 1 & 1 & 1 & 0 & 0 & 0 \\ 1 & 1 & 0 & 0 & 1 & 1 & 0 \\ 1 & 0 & 1 & 0 & 1 & 0 & 1 \end{bmatrix}$

(2) *If* $R = \begin{bmatrix} b_7 & b_6 & b_5 & b_4 & b_3 & b_2 & b_1 \end{bmatrix}$, *then*

(3) $S = HR^T = \begin{bmatrix} b_7 + b_6 + b_5 + b_4 + 0 + 0 + 0 \\ b_7 + b_6 + 0 + 0 + b_3 + b_2 + 0 \\ b_7 + 0 + b_5 + 0 + b_3 + 0 + b_1 \end{bmatrix}$

If there are no errors the syndrome S must equal zero. This means the sum of each row has to equal 0. This is why the rows of S are defined as parity equations whose no error sum is 0 (this is not obvious, see equations 6).

---

[1] Hamming, R.W. 1950. Error Detecting and Error Correcting Codes. *Bell System Technical Journal* Vol. 29 No. 1 (April): 147-160.

Hamming deduced that bits $b_1$, $b_2$, and $b_4$ had to be the parity bits, because each appears *in only one row* (equation 3) so that each is in only one parity equation (6). He labeled received bits $b_1$, $b_2$, $b_4$ as parity bits $r_1$, $r_2$, $r_3$, and bits $b_3$, $b_5$, $b_6$, $b_7$ as message bits $k_1$, $k_2$, $k_3$, $k_4$. Let C be the code word.

**Parity equations** Compare code word C to received word R where we rewrite the b's as message bits k and parity bits r. In this way Hamming related the k message bits to code word bits $c_7$, $c_6$, $c_5$, $c_3$, and the r parity bits to code word bits $c_4$, $c_2$, $c_1$.

(4) $C = c_7 \quad c_6 \quad c_5 \quad c_4 \quad c_3 \quad c_2 \quad c_1$

(5) $R = b_7 \quad b_6 \quad b_5 \quad b_4 \quad b_3 \quad b_2 \quad b_1$

$\quad R = k_4 \quad k_3 \quad k_2 \quad r_3 \quad k_1 \quad r_2 \quad r_1$ *(no errors)*

$(6a)\, b_7 + b_6 + b_5 + b_4 = 0 \qquad k_4 + k_3 + k_2 + r_3 = 0$

$(6b)\, b_7 + b_6 + b_3 + b_2 = 0 \quad \Rightarrow \quad k_4 + k_3 + k_1 + r_2 = 0$

$(6c)\, b_7 + b_5 + b_3 + b_1 = 0 \qquad k_4 + k_2 + k_1 + r_1 = 0$

**BCH** In 1960 BCH[2] (Bose, Ray-Chaudhuri, Hocquenghem) replaced Hamming's r with BCH parameter m to allow for a disconnect of n from r by *defining $n=2^m-1$ and $r=mt$* (see the Example page 179). BCH showed that for the (n,k) binary block codes t is the number of errors the decoder can correct (if t=1, then $r=mt=m$, $n=2^r-1$ and we are back with Hamming).

The syndrome matrix now has *t groups of m rows* that equal a total of $r=mt$ rows. Consequently each column of H has t groups of m bits ($r=mt$). In a Hamming code t=1, and the H matrix has one group of $mt=m=r$ rows. Decoding requires syndrome logic, an error bit numbers calculator, and xor correction logic (Figure 819).

**Figure 819 Encoding and Decoding a word**

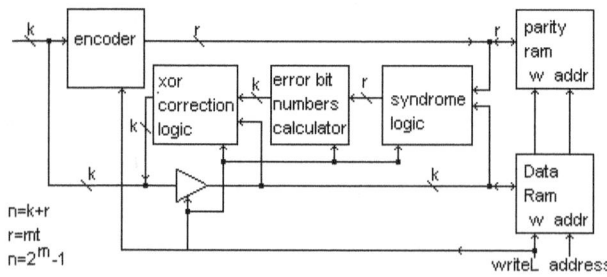

---

[2] Bose and Ray-Chaudhuri. 1960. On A Class of Error Correcting Binary Group Codes. *Information and Control* Vol. 3 (March): p68-79.

BCH achieved this result by applying Galois finite field theory GF($2^m$). See page 146 for a sample. In turn this created the need to apply other complex mathematical theories that brings us to code design.

***Example*** t=2 error correcting BCH code for 32 bit message words
*BCH Theorem 3 If n = $2^m-1$, then there is a t error correcting (n,k) binary group code where $k = 2^m - 1 - R(mt) \geq 2^m - 1 - mt$*

If k=$2^5$=32 and r>0, then n=k+r >32. Now $2^6$ > (n>32) > $2^5$. This means m=6 and $n = 2^m - 1 = 2^6 - 1 = 63$, $r = mt = 6 \times 2 = 12$, $k = n - r = 63 - 12 = 51$. Computing words are some power of two, 51 is not a power of 2, nor is it the desired 32. We achieve k=32 when we apply BCH Corollary 1 that disconnects from the constraint that n=$2^m$–1.

*BCH Corollary 1 The existence of a t error correcting (n,k) binary group code implies the existence of a t error correcting (n′, k′)=(n–c, k–c) binary group code, 0<c<k.*

In a 32 bit word computing system k′=32, so that c = k–k′ = 51–32 = 19.

$r = n - k = 63 - 51 = 12 = mt = 6 \times 2$ $\Rightarrow (n,k) = (63,51)$ *linear code*
*If* $c = 19$, *then* $n - c = 44, k - c = 32, r = 12$ $\Rightarrow (n,k) = (44,32)$ *linear code*

=====

*The mathematical structure of Code Design Theory is constructed from*
1. P. Fermat's (1601-1665) Theorem: If p is prime and n is a positive integer, then p divides $n^p$–n, or $n^{p-1} = 1$ mod p or $n^p \equiv n$ mod p.

2. E. Galois' (1811-1832) finite field GF($2^m$).

3. A. Vandermonde's (1735-1796) matrices. The 2t×2t Vandermonde matrix has rank 2t.

4. K. F. Gauss' (1777-1855) concept of congruence applied to operations + – × ÷ on polynomials so that binary strings (n-tuples) can be manipulated by interpreting their 0's and 1's as coefficients of terms of polynomials.

5. I. Newton's (1642-1727) Theorem which deals with sums of powers of roots of a polynomial.

=====

***Code design*** Code design[1] is a very complex subject using very complex mathematics. The ability to design codes requires a significant effort to acquire knowledge and understanding of the mathematics on which code design is based. With the knowledge in hand you can proceed to devise a process, which turns out to be straightforward. Here is such a process.

First know this: *You still have to invent circuits to implement the code you designed* (Figures 818, 819).

*Derive the software: 1. generating polynomial g(x), 2. G matrix encoder for a systematic code, 3. H matrix decoder.*
*Design the hardware: 1. Encoder, 2. Decoder consisting of syndrome logic, error bit lookup logic, and Xor correction logic.*

The following steps are a process that design an (n.k)= (44,32) error correcting BCH code where n = 44, k=32, r=12, m=6, t=2.

═══════════════════════════════════════════

***Example Use a Galois Finite Field GF(2) and extension field GF($2^m$)***
Let m = 3 so that the order of $\alpha$ is $2^m-1=7$ (Polynomials mod 2 page 146).
$$x^7 + 1 = (x+1)(x^3 + x + 1)(x^3 + x^2 + 1) = m_0(x)m_1(x)m_2(x)$$
If $\alpha$ is a root of $x^7 + 1$, then $\alpha^7 + 1 = 0 \Rightarrow \alpha^7 + 1 + 1 = 0 + 1 \Rightarrow \alpha^7 = 1$

*Reduce by $\alpha$* An easy way to reduce polynomials to degree n−1=7 or less is to use the fact $\alpha$ is a root of one of the factors $x^7+1=0$. We can select any factor. If $f(x) = x^3 + x + 1$, then $f(\alpha) = \alpha^3 + \alpha + 1 = 0$ *and so* $\alpha^3 = \alpha + 1$
Then we create all powers of $\alpha$.

| mark | polynomial | | polynomial | 3-tuple |
|---|---|---|---|---|
| $\alpha^0$ | 1 | | $0\alpha^2 + 0\alpha + 1 \cdot 1$ | 001 |
| $\alpha^1$ | $\alpha$ | | $0\alpha^2 + 1\alpha + 0 \cdot 1$ | 010 |
| $\alpha^2$ | $\alpha^2$ | | $1\alpha^2 + 0\alpha + 0 \cdot 1$ | 100 |
| $\alpha^3$ | $\alpha + 1$ | | $0\alpha^2 + 1\alpha + 1 \cdot 1$ | 011 |
| $\alpha^4$ | $\alpha\alpha^3 = \alpha(\alpha+1) = \alpha^2 + \alpha$ | | $1\alpha^2 + 1\alpha + 0 \cdot 1$ | 110 |
| $\alpha^5$ | $\alpha^3\alpha^2 = (1+\alpha)\alpha^2 = \alpha^2 + \alpha^3 = \alpha^2 + \alpha + 1$ | | $1\alpha^2 + 1\alpha + 1 \cdot 1$ | 111 |
| $\alpha^6$ | $\alpha^3\alpha^3 = (\alpha+1)^2 = \alpha^2 + 0\alpha + 1 = \alpha^2 + 1$ | | $1\alpha^2 + 0\alpha + 1 \cdot 1$ | 101 |
| $\alpha^7$ | 1 | | 1 | 001 |

$x^7+1=0$ has 7 roots, which are $\alpha_0$ to $\alpha_6$.
*The powers of $\alpha$ can be computed for any m.*

═══════════════════════════════════════════

---
[1] Nicholas. L. Pappas, *Error Correction Code Design*

*1. Generating polynomial g(x)* When m=6 the powers of $\alpha$ range from 0 to 63. Generating polynomial g(x) has to be the product of some factors of $x^{63}+1$. On page 145 find the reason why the degree of g(x) equals r. The degree of $g(x)=m_3(x)\,m_1(x)$ is 12, which equals r=mt=6×2 in this example.

$$x^{63}+1 = m_0(x)m_1(x)m_2(x)m_3(x)m_4(x)m_5(x)m_6(x)$$
$$m_7(x)m_8(x)m_9(x)m_{10}(x)m_{11}(x)$$

| factor | roots | minimal polynomial |
|---|---|---|
| $x+1$ | $\alpha^0$ | $m_0(x)$ |
| $x^6+x^1+1$ | $\alpha^1,\alpha^2,\alpha^4,\alpha^8,\alpha^{16},\alpha^{32}$ | $m_1(x)$ |
| $x^6+x^4+x^2+x^1+1$ | $\alpha^3,\alpha^6,\alpha^{12},\alpha^{24},\alpha^{48},\alpha^{33}$ | $m_3(x)$ |

*others not shown*

$$g(x) = m_3(x)m_1(x)$$
$$g(x) = (x^6+x^4+x^2+x^1+1)(x^6+x^1+1)$$
$$= x^{12}+x^{10}+x^8+x^5+x^4+x^3+1 = 1010100111001$$

*2. G matrix encoder* The process starts by multiplying the message word k(x) by $x^r=x^{12}$. This forms a polynomial of degree n=k+r (44=32+12). The message bits are now in bit-positions corresponding to $x^{43}$ to $x^{12}$, and zeros are in bit-positions corresponding to $x^{11}$ to $x^0$. Dividing message word $x^{12}m(x)$ by generating polynomial g(x) produces remainders r(x).

> A bonus we discovered is that g(x) dividing the k bit message word 100...0000 enhanced by r zero parity bits produces *all* the parity equations r(x). Here is k=32 and r=12.

$$g(x){\overline{\smash{\big)}\,x^{12}k(x)}}\overset{q(x)}{\phantom{x}}$$

```
                 00.............1010001...
  1010100111001 )1000000000000000000000...
                 1010100111001
            r₀   010100111001
                 0000000000000
            r₁   101001110010
                 1010100111001
            r₂   000111011101
                 0000000000000
            r₃    001110111010   and so forth
```

When k=32, r=12 the G matrix is formed from a 32×32 *unit* matrix $I_{32\times32}$ and 32×12 parity matrix P. We selected the *unit* matrix $I_{32\times32}$, because its rows are valid 32-bit message words 10...0000, 01...0000, ... , 00...0100, 00...0010, and 00...0001. And, $I_{32\times32}$ is a simple as it gets and has rank 32. The parity matrix P has 32 rows of 12 bit $r_j$ numbers representing the r parity equations.

$$G_{32\times44} = \begin{bmatrix} I_{k\times k} & P_{k\times r} \end{bmatrix} = \begin{bmatrix} I_{32\times32} & P_{32rows\times12columns} \end{bmatrix}$$

Codeword $C_{1\times44} = K_{1\times32}G_{32\times44}$ where message K and codeword C are

$$K = \begin{bmatrix} k_{31} & k_{30} & k_{29} & \ldots & k_2 & k_1 & k_0 \end{bmatrix}$$ and

$$C = \begin{bmatrix} k_{31} & k_{30} & k_{29} & \ldots & k_2 & k_1 & k_0 & r_{11} & r_{10} & r_9 & \ldots & r_3 & r_2 & r_1 & r_0 \end{bmatrix}$$

Observe that code word c(x)=1100...000 is the sum of G matrix code words 1000...000 and 0100...000. In this way all code words can be produced. In other words the code word c(x) is formed by adding the remainder r(x) to $x^{12}k(x)$. This is the encoding process.

*3. H matrix decoder* The H Matrix has mt = 6×2 = 12 rows and n = 44 columns). One can demonstrate that the H matrix is derived from the G matrix.

$$G_{32\times44} = \begin{bmatrix} I_{k\times k} & P_{k\times r} \end{bmatrix} = \begin{bmatrix} I_{32\times32} & P_{32rows\times12columns} \end{bmatrix} \quad \text{32 rows by 44 columns}$$

$$H_{12\times44} = \begin{bmatrix} P^T_{k\times r} & I_{r\times r} \end{bmatrix} = \begin{bmatrix} P^T_{32\times12} & I_{12\times12} \end{bmatrix} \quad \text{12 rows by 44 columns}$$

*Opinion and word of caution*: if you pursue this subject you will run into a lot of talk and mathematics, which is interesting but irrelevant to actual code design. We spent a lot of time separating the wheat from the chaff.

---

*Problem* 814 Code can only correct one error A received code word is $b_7b_6b_5b_4b_3b_2b_1$. The parity equations produce the syndrome column number $s_2s_1s_0$. The parity equations are `

$$s_2 = b_4 + b_5 + b_6 + b_7 \qquad s_1 = b_2 + b_3 + b_6 + b_7 \qquad s_0 = b_1 + b_3 + b_5 + b_7$$

A received code word is 1010101. Show that the syndrome $S=HR^T$ is 000 (no errors). The next received code word is 1000101 that has an error in bit $b_5$. Show that the syndrome is 101. The next received code word is 1001101 that has errors in bits $b_5$ and $b_4$. Show that the syndrome is 001. This syndrome reports $b_1$ as an error. This is a decoding failure. The reason is that this code can only correct one error.

---

# Review 8

Everything useful is operated by a state machine, and every state machine uses memory. The memory may be just a one bit register, or $2^{25}$ bits supporting a central processing unit (cpu) that is one form of state machine. Memory stores words for use on demand.

A memory chip is basically an assembly of an address decoder, an array of one bit storage cells grouped as k bit words where k is a power of 2, and a two way input/output circuit with k data lines known as d/q (in/out).

A realistic view of the memory chip is this. There is the address side where *n* address lines select a word of k one bit cells from the array of one bit storage cells and connect the cell's 1,0 nodes to *internal* bit lines. There is the i/o side where d/q data lines transfer an external word to write to, or an internal word to read from, the *internal* bit lines of the selected k one bit cells. A reasonable approximation is that selection and i/o processes each use up one half the access time.

A *CPU Bus Protocol* provides a realistic context that shows how access time affects performance. Almost all computers use a bus protocol where a read or write transfer cycle requires a minimum of three cpu bus clock periods. Know that there are many bus protocols.

Every digital system includes a state machine controlling a *data path*. The memory system data path is defined by the data storage requirement. A basic memory data path has an SRAM (static ram) or DRAM (dynamic ram) memory chip array of one or more chips, an address decoder, and registers to store address and data. System implementation has many forms.

Cost of memory chips is a, if not the, major reason for using memory hierarchies. A memory hierarchy organization is based on the fact memory chip cost per bit is inversely proportional to bit access time. Cost per bit increases dramatically as access time decreases.

The purpose of an error detection and correction system is to detect and correct errors. This is why bits are numbered from 1 to n according to their position in the n bit code word. If we know the positions of incorrect bits, then we can correct them because we know which bus wires carry the bits.

# 9 Microprogrammed State Machines

Maurice Wilkes invented microprogramming about 1950. He showed[1,2] how any state machine can be can physically realized by microprogramming. He recognized that logic design naturally separates into two parts: the control state machine, and the controlled data path. The ASM had not been invented at the time he proposed designing a computing

**Figure 901 Wilkes' State**

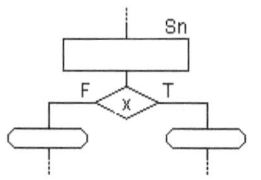

machine that executed a sequence of states where each state had only *zero or one* input variable associated with it, and perhaps conditional outputs (Figure 901). The two equations representing this state are implemented as a Wilkes' gate (Figures 901, 902).

$$F = x' \cdot S_n \qquad\qquad T = x \cdot S_n$$

**Figure 902 Wilkes' Gate with *inputs x and state $S_n$***

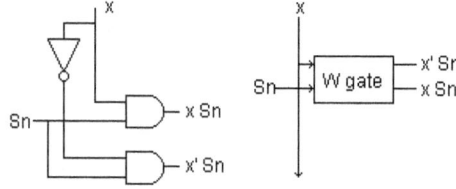

Wilkes' ingenious idea simplified ASM state machine circuit designs by using a matrix of diodes format (Figure 904). One way to describe his method is by an example. The ASM design process (Section 5.2 page 97) produces the truth table for this ASM chart (Figure 903).

**Figure 903 ASM Chart**

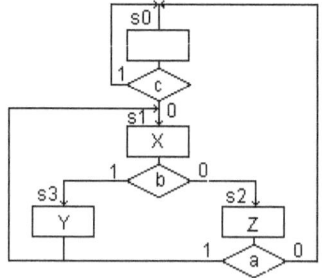

---

[1] Maurice V. Wilkes The genesis of microprogramming reprinted in IEEE Annals of the History of Computing, pp. 116-126, 1986, ISSN 1058-6180
[2] Wilkes and Stringer 1953 follow-up paper reprinted as Chapter 11 of Computer Structures, Principles, and Examples, ISBN 0070 573 026

Truth table for ASM chart (Figure 903)

| Present State | | | | Next State | Outputs | | |
|---|---|---|---|---|---|---|---|
| $S_n\, q_1 q_0$ | c | B | a | $S_n\, q_1^{+} q_0^{+}$ | X | Y | Z |
| $S_0$ 00 | 0 | - | - | $S_1$ 01 | 0 | 0 | 0 |
| $S_0$ 00 | 1 | - | - | $S_0$ 00 | 0 | 0 | 0 |
| $S_1$ 01 | - | 0 | - | $S_2$ 10 | 1 | 0 | 0 |
| $S_1$ 01 | - | 1 | - | $S_3$ 11 | 1 | 0 | 0 |
| $S_2$ 10 | - | - | 0 | $S_0$ 00 | 0 | 0 | 1 |
| $S_2$ 10 | - | - | 1 | $S_1$ 01 | 0 | 0 | 1 |
| $S_3$ 11 | - | - | - | $S_1$ 01 | 0 | 1 | 0 |

The next state equations are

$$q_1^{+} = S_1 \qquad q_0^{+} = c'S_0 + bS_1 + aS_2 + S_3$$

The output equations are

$$X = (b' + b)S_1 \qquad Y = S_3 \qquad Z = (a' + a)S_2$$

Inputs c, b, a are paired with states $S_0$, $S_1$, $S_2$ in Wilkes' gates to define 6 horizontal matrix rows (Figure 904). Unpaired state $S_3$ defines the seventh matrix row. Rows are connected to vertical output lines via diodes. State flip-flops $q_1 q_0$ and a state decoder complete the circuit.

**Figure 904 Wilkes' State Machine for the ASM (Figure 903)**

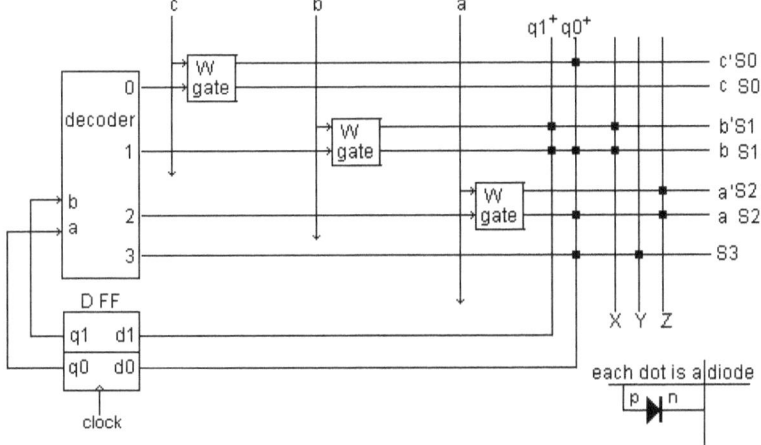

Wilkes used diodes to build his machine, because in those days diodes were used to implement AND and OR gates. Each vertical line is an OR gate output. When any decoder or W gate output goes H (true) the vertical line goes H.

Digital Design

The diode matrix is a hardwired memory addressed by a 5 bit address $q_1 q_0 cba$ reading out 5 bit data words $q_1^+ q_0^+ XYZ$ (the two 5's is a coincidence).

As the years flew by the four states of the diode matrix were perceived as four *microinstructions* that *microprogrammed* the state machine. This bring us to a modern view.

*A modern computer design view* Inputs a, b, c modify the sequence of states as shown in modern ASM format (Figure 903). Wilkes' diode matrix (Figure 904) implements the state machine sequence of states, and state actions. The Wilkes' state machine *program* is fixed, because its set of actions cannot be changed unless the wiring is changed.

Over time programmable state machines were desired. And so Wilkes' diode matrix evolved into a stored computer program that is a list of computer instructions stored in *sequential* random access read/write memory words. Want to change the actions? Store a new program.

What do we mean when we say *store a new program?* We mean store binary words in read/write memory that are intended to be analogous to a program stored in a diode matrix. We store in read/write memory *words* that are instructions coded as a series of binary digits (bits). Then we read them out in sequence one at a time and execute them. Branches may change the sequence. Anticipating the upcoming computer design we start by defining cpu instructions. Here are some of them.

| Instruction | Action |
|-------------|--------|
| ADD ra rb rd | $rd \leftarrow rb$ add ra |
| ADDI n rb rd | $rd \leftarrow n$ add rb |
| BEQ n | if $Z=1$, then pc $\leftarrow pc+2+2n$, else pc $\leftarrow pc+2$ |
| CONSTH m rd | $rd \leftarrow m << 6$ (shift m left 6 times, 0 fill) |
| IN rd rb | $rb \leftarrow rd$ (fake uI for clarity - use LDW) |
| OUT rb rd | $rd \leftarrow rb$ (fake uI for clarity - use STW) |
| LDW *rb(n) rd | $rd \leftarrow M[rb+n]$ (LDW is Load Word) |
| STW rd *rb(n) | $M[rb+n] \leftarrow rd$ (STW is Store Word) |

In our elementary computer user instructions are represented by 16 bit words stored in 16 bit memory words. *Fields* in any 16 bit user instruction word define parameters (opcodes, registers, numbers m and n). Three bit fields limit the number of registers to 8. The number of bits in m and n fill out the uI words to 16 bits.

| 15 | | 12 | 9 | 6 | 3 | 0 |
|---|---|---|---|---|---|---|
| 4 bits | | 3 | 3 | 3 | 3 | |
| opcode | | rd | rb | ra | not used | |

$ADD\ ra\ rb\ rd \rightarrow ADD\ r_5\ r_7\ r_2 \rightarrow 0000\ 010\ 111\ 101\ 000$

| 4 bits | 3 | 3 | 6 |
|---|---|---|---|
| opcode | rd | rb | n (signed constant) |

$ADDI\ n\ rb\ rd \rightarrow ADDI\ 22\ r_1\ r_6 \rightarrow 0001\ 110\ 001\ 010110$

| 4 bits | 3 | 9 |
|---|---|---|
| opcode | rd | m (signed constant) |

$CONSTH\ m\ rd \rightarrow CONSTH\ -41\ r_3 \rightarrow 0011\ 011\ 111010111$
$-41$ equals 2's complement of $+41$ that equals 1's complement of 41 plus 1
$+41_{10}=000\ 101\ 001_2$ so that $-41_{10}=111\ 010\ 110_2 +1_2=111\ 010\ 111_2$

Variables are stored in registers. A zero is stored in register $r_0$ forever by wiring $r_0$ as a zero. Use *ADDI* to store X, Y, W in registers $r_4$, $r_5$, $r_6$. Use *ADD* to set status bit Z, and *BEQ* to go to next state. Emphasis: Z is a status bit. Here is the ASM in code.

**Figure 903 ASM Chart**

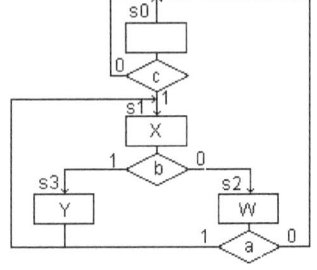

| | | |
|---|---|---|
| ADDI X $r_0$ $r_4$ | ($r_4 \leftarrow r_0+X$) | |
| ADDI Y $r_0$ $r_5$ | | |
| ADDI W $r_0$ $r_6$ | | |
| $S_0$ IN kbd $r_3$ | (input c from kbd) | |
| ADD $r_0$ $r_3$ $r_3$ | (if c=0, then $r_3$=0 and status logic sets Z=1) | |
| BEQ $S_0$ | (if c=0, then go to $S_0$ else go to $S_1$) | |
| $S_1$ OUT $r_4$ $R_X$ | (store X in output register $R_X$) | |
| IN kbd $r_3$ | (input b from kbd) | |
| ADD $r_0$ $r_3$ $r_3$ | (if b=0, then $r_3$=0 and status logic sets Z=1) | |
| BEQ $S_2$ | (if b=0, then go to $S_2$ else go to $S_3$) | |
| $S_3$ OUT $r_5$ $R_Y$ | (store Y in output register $R_Y$) | |
| ADD $r_0$ $r_0$ $r_0$ | (Z=1) | |
| BEQ $S_1$ | | |
| $S_2$ OUT $r_6$ $R_W$ | (store W in output register $R_W$) | |
| IN kbd $r_3$ | (input a from kbd) | |
| ADD $r_0$ $r_3$ $r_3$ | (if a=0, then $r_3$=0 and status logic sets Z=1) | |
| BEQ $S_0$ | (if a=0, then go to $S_0$ else go to $S_4$) | |
| $S_4$ ADD $r_0$ $r_0$ $r_0$ | (Z=1) | |
| BEQ $S_1$ | ($S_4$ is a fake state to be able to go to $S_1$) | |

# 10 CISC 8

The goal is the design of a *complex instruction set computer* (CISC). We will show how the instruction set specifies the hardware. We will *only* design the datapath and the control. We are in luck, because the hardware design process is straightforward. However the demands on your skills will be significant. And, as we deal with the too many trees we must not lose sight of the forest.

Before Wilkes (Chapter 9), computers used essentially random logic. Maurice Wilkes's method of implementing microinstructions that executed *user instructions* (*uI*) transformed computer design and construction from random logic, with its unavoidable errors, to orderly circuits that have *systematic layouts*[1].

The purpose of a computer is to execute *programs*. A program is a list of instructions *the user writes* that the computer can execute by design. What follows is *not* about programming, nor how to design and select a set of user instructions (uI).

A commercially useful computer can have 100's of user instructions (a code forest with many, many trees). We start by selecting a scaled down user instruction (uI) set, which is the basis of the design. A scaled down uI set avoids cluttering the design with too many trees in the forest, so that we can focus on the ideas enabling the design. Nevertheless this is a reasonably complex design.

Today programs are written and stored as n bit words in a permanent memory such as a disk drive. In order to focus on the design we simply say the program is transferred to the computer memory in the form of n bit words.

Any program as written consists of instructions from the computer's instruction set. The first design task is to define the user instruction set – the uI (Section 10.1). The uI set has three major subsets, which are arithmetic logic instructions (alu), branch instructions that enable complex program procedures (br), and memory load (load) and store (store) uI.

---

[1] Nicholas L. Pappas, *CMOS Circuit Design – Analog, Digital, IC Layout*

The uI can access memory in several ways known as address modes (Section 10.2).

Significant contributors to program capability are the branch uI. The branch uI are controlled by status bits that represent the result of an arithmetic operation (Section 10.3).

Storing a uI in memory as an n bit word requires that the bits in the words have specific meaning (Section 10.4).

Unfortunately the n bit word representing a uI cannot be executed directly with systematic hardware (this is not obvious). This is why each uI is replaced in the program by a list of microinstructions (mI) as shown in Section 10.5.

> A uI can be executed directly by special hardware circuits in one clock cycle. Then the computer is a RISC computer - a reduced instruction set computer, which we do not design here.

Execution of a user instruction takes place in the *user data path* (*uDP*). We show how to transform the set of uI into a cpu control and a user datapath uDP, which can execute one mI at a time (Section 10.6)

The user data path uDP is controlled by the *cpu control* (Section 10.7). The user data path is a set of building blocks interconnected by data buses.

*Emphasis*: the uDP and cpu control circuits are specified by the CISC user instruction (uI) set. However the number of bits in a word is not.

We select 8 bits for memory word size to simplify matters. The 8 bit decision sets the uDP registers' size to 8 bits, and buses to 8 wires each

> *The user data path uDP is configured by control lines asserted by the cpu control outputs so that the uDP executes mI one at a time.*

# 10.1 Instruction Set

Included in the user instruction set are operators on numbers, program control branches, and memory read and write operations to fetch and store data and uI.

*Operations on numbers* We can add and subtract numbers, do logical operations NOT, AND, OR, XOR, use subtraction to compare numbers, clear a register to zero, and shift a number left and right (multiply and divide by 2), etc.

*Program Control* An instruction performs its task, and in the process may change the value of one or more status bits. The next branch instruction that appears will base its *branch or not-to-branch* decision on the values of those status bits.

*Memory read and write* As any program proceeds instructions, numbers, data have to be fetched from memory and written to memory.

*Register Transfer Language* Register transfer language (RTL) symbols are used in Table 1001 to define the function of each uI.

The left arrow ← is the replacement symbol. The contents of register P are replaced by the contents of register Q in P ← Q.

Keep in mind that any *register* is a stand alone memory location.

The symbol for *the contents of* memory location x is M[x] where x is the address. Various memory transfer statements are

M[x] ← $r_A$    (contents of register $r_A$ stored at memory address x)
$r_A$ ← M[x]    (contents of memory address x to register $r_A$)

When the contents of registers and memory words are modified, statements such as pc ← pc+1, rd ← rd−1, or rd ← rd+rt+c may not be clear. *The plus and minus signs refer to adding and subtracting the contents of registers.*

**TABLE 1001** CISC-8 uI set

| Operators: | Action (These uI clear and set N C Z V status bits.) |
|---|---|
| ADD M rd | rd ← rd + M (+ means plus) (rd is rA or rX) |
| | (M is *iw* or content of *an address*) |
| SUB M rd | rd ← rd − M (− means minus) |
| INC rd | rd ← rd+1 |
| DEC rd | rd ← rd − 1 |
| | |
| NOT rd | rd ← rd' |
| AND M rd | rd ← rd AND M |
| OR M rd | rd ← rd OR M |
| XOR M rd | rd ← rd XOR M |
| | |
| SLL rd | rd ← rd << rd (shift 1 bit left, zero fill) |
| SRL rd | rd ← rd >> rd (shift 1 bit right, zero fill) |
| ROTL rd | rd ← rd << rd (msb to lsb for each 1 bit shift) |
| ROTR rd | rd ← rd >> rd (lsb to msb for each 1 bit shift) |
| | |
| BRcc n | if cc true then pc ← pc+n else pc ← pc+1 |
| | |
| LOAD M rd | rd ← M (M is *iw* or content of *an address*) |
| STORE z M | M ← z  (z is $r_A$ or $r_X$) (M is content of *an address*) |

**TABLE 1002** CISC-8 uI address modes (explained in 10.2)

| type | Address Modes | ea examples |
|---|---|---|
| | Inherent | SLL rd |
| iw | Immediate | ADD iw rd |
| addr | Direct - add contents at address | ADD addr rd |
| *rx(n) | Register indirect with signed offset n | ADD *$r_X$(n) $r_A$ |
| | (index mode) | ADD $r_A$ *$r_X$(n) |

Note: Index mode only uses registers $r_A$, $r_X$ as shown.

## 10.2 Address Modes

Address modes define how an architecture calculates the address of the data to be accessed. Address modes contribute to the complexity of CISC machines. Each CISC-8 uI operand uses any 1 of 4 address modes. A two operand uI uses address modes in 4×(4−1) combinations, because, as will become clear, the immediate mode cannot be a destination. On the other hand address modes reduce the number of uI needed in a program. Address modes make user programming easier. The price paid is learning how to use the address modes. A greater price is learning what address modes have been implemented for which uI. *Know that not all address modes are implemented for all uI in most computers for practical reasons.*

The data is in memory. This is basically why discussions of instruction sets are preoccupied with the various processes of loading data from memory and storing data in memory. Please keep in mind that address modes are more readily understood when you ask

1. Where is the address stored?
2. Where is the data stored?

*Effective address (ea)* is a useful notation for the memory location actually accessed by a uI. The concept of effective address provides greater clarity when discussing memory references. This is because the actual address calculations are not germane to many discussions.

A uI may have a program constant associated with it. Literal is a generic name for program constants. The rule we adopt is this.

> *Literals are placed in memory locations following the memory location of the uI code word associated with the literal.*

A simple, yet useful, model for *reading a memory* is that a memory is a circuit with storage registers whose input is an address and whose output is a data word on a data bus.

There is one data word stored at each address. Furthermore the address can access only one data word at a time. The memory is a bank of *registers* that can be accessed only one register at a time.

The simple model for *writing memory* is that a memory is a circuit with registers whose inputs are an address and a data bus. The data word on the bus is stored in the addressed memory register.

Programs are lists of user instructions uI and the associated literals stored in memory. One uI or one literal is stored in each memory register allocated to the program. Why the following addressing modes exist is a programming issue we do not address. The address modes are listed in Table 1002 page 191, and examples are shown in Figure 1001.

This is important: we *arbitrarily* chose to have literals such as iw and n follow the uI code word in memory (Figures 1001, 1002, 1003. *If the uI is at address m, then its literal is stored at address m+1.*

### Figure 1001 Address Mode Examples

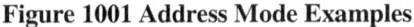

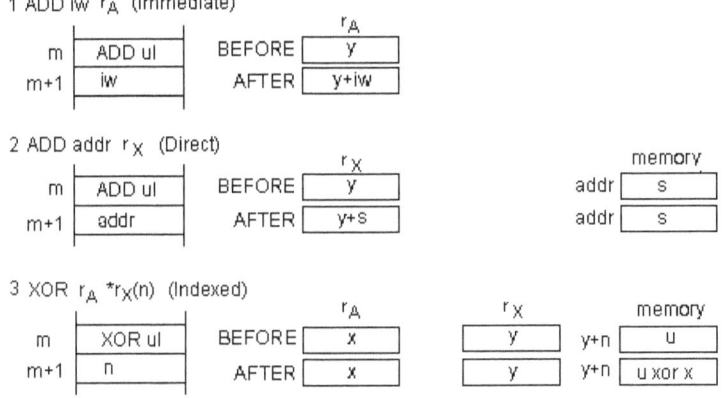

### Figure 1002 Cisc-8 Programming Model

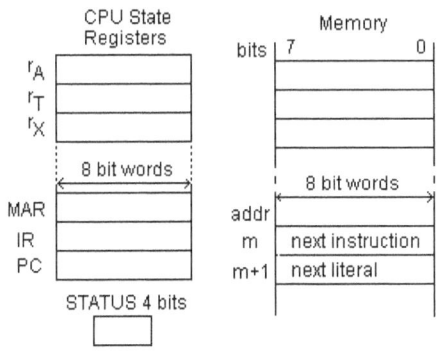

### Figure 1003 uI Address Modes

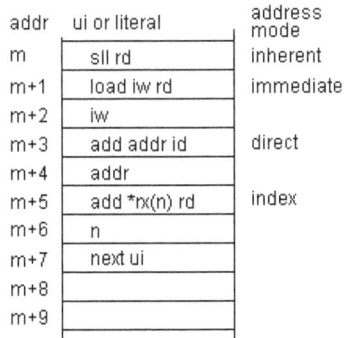

Digital Design

The address mode descriptions:

*Inherent: SLL rd* Inherent addressing implies unary operations such as shifting a word in a register.

*Immediate addressing: ADD iw rd* The easiest way to reach the data is to make it part of the instruction. We can do this in two ways: incorporate the *immediate word iw* into the uI code word or let iw be a literal following the uI code word in memory. Our 8 bit words require us to make iw a literal following the uI code word in memory.

The iw is one of the uI operands. Literal iw can only be source data. Can you store data in an iw? Not really. This is why there are 4−1 choices for uI operands that are destinations, and 4 choices for sources. iw cannot be a destination.

*Direct addressing: ADD addr rd* Another easy way to reach data makes one operand the data address. In our 8 bit computer the literal addr uses one 8 bit word to represent an 8 bit address. The literal *addr* follows the uI code word in memory. Note that addr can be the address of the source data or the destination data (as in ADD rs addr). The uI code word is at address m, the addr word is at address m+1.

*Register Indirect with signed displacement n: XOR rs \*$r_X(n)$* If the content of $r_X$ written as $(r_X)$ is y, then the destination data address is $(r_X)+n = y+n$. [A source of confusion is *incorrectly* writing this as $r_X+n$. However, the *incorrect* writing is convenient, which is why many writers, including yours truly, use it.] The sum, contents y of $r_X$ plus n, is an address pointing to the data. The literal n is a word. Literal n follows the uI code word in memory. The uI code word is at address m, and the *index* n is at address m+1. The *contents* of $r_X$ is sometimes called a base address to which the index n is added.

# 10.3 Status Logic and Condition Codes

Almost all programs need to test data values and change the sequence of uI executed according to the test results. This is *program control*, which is implemented by branches to other parts of the program under specified conditions. These conditions are known as condition codes *cc*, which are the basis for action by branch uI. The uI branch decisions are determined by the condition code equations (Table 1004 page 197), which use the N C Z V status bits as variables. Status bit values are produced by results of operations executed by operator uI.

This means a branch decision requires two uI. An operator uI executes prior to the branch uI and determines status bit values. The second uI, the branch uI, uses the status bit values in a *cc* equation to make a branch decision. The program branches to another place in the program if the equation is true. If the equation is false the program proceeds to execute the next in line uI (page 187).

*Status bits* The status register has 4 bits with names N C Z V (Table 1003). The meaning attached to these bits follows. Msb is the most significant bit.

*TABLE 1003* Status Bits N C Z V defined

| Bit | Name | Action by operator uI |
|-----|------|------------------------|
| N | sign bit | Set to 1 if result msb is 1 |
| | | Set to 0 if result msb is 0 |
| C | carry bit | Set to 1 if result generates a carry |
| | | Set to 0 if result does not generate a carry |
| Z | zero bit | Set to 1 if result is 0 |
| | | Set to 0 if result is not 0 |
| V | overflow bit | Set to 1 if result overflows |
| | | Set to 0 if result does not overflow |

*Problem* 1001 Add $200_{10}$ and $183_{10}$. Which NCZV bits are set to 1?

*Problem* 1002 x=$200_{10}$, y=$183_{10}$. Do unsigned compare. Which NCZV bits are set to 1?

*Problem* 1003 x=$200_{10}$, y=$183_{10}$. Do arithmetic (signed) compare. Which NCZV bits are set to 1?

Status bit values are determined by (hardware) equations. Only the programmer knows whether or not a bit value is valid. For example if numbers are unsigned, then N is irrelevant.

**Sign N** The sign bit value is based on the msb value. The hardware is a robot calculating N=msb.

**Carry C** The carry bit from the msb has meaning only when the numbers are interpreted as unsigned. Carry bit C reports the carry from the msb for any ALU operation. The carry produced by combinations of positive and negative numbers is discussed in the upcoming paragraphs on Overflow.

**Zero Z** Z is set to 1 when the result is zero, because 1 represents true.

**Overflow V** Overflow occurs when operations on signed numbers produce a result out of the signed number range. To find a way to detect overflow let us examine the addition of two positive or two negative numbers. By the way, an overflow does not occur when numbers of different signs are added (why?). The problem arises in a system of + and − numbers where *the msb is the sign bit.*

| carries | 1000 | | 1100 | | 1110 | |
|---|---|---|---|---|---|---|
| $x$ | 1001 | $-7$ | 1100 | $-4$ | 1010 | $-6$ |
| $y$ | $+1010$ | $-6$ | $+1110$ | $-2$ | $+1110$ | $-2$ |
| $f = x + y$ | 0011 | $+3$ | 1010 | $-6$ | 1000 | $-8$ |
| overflow? | yes | | no | | no | |

| carries | 0110 | | 0000 | | 0110 | |
|---|---|---|---|---|---|---|
| $x$ | 0111 | $+7$ | 0100 | $+4$ | 0110 | $+6$ |
| $y$ | $+0110$ | $+6$ | $+0010$ | $+2$ | $+0010$ | $+2$ |
| $f = x + y$ | 1101 | $+13$ | 0110 | $+6$ | 1000 | $-8$ |
| overflow? | yes | | no | | yes | |

Observation: On overflow the x and y msb bits are always equal AND the f msb bit is *different* from the x and y msb bits. Therefore the equation for V is $V = fy'x' + f'yx$     *where x, y, f are msb bits*

**Figure 1004 mcc Logic**

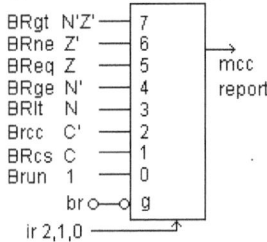

*mcc status logic* The mcc logic is in the user data path (Figure 1006 page 207). The mcc_report is an input to the cpu control logic (Figure 1010 page 209 ).

The NCZV bits are derived from alu and shifter outputs (Figure 1005). The NCZV bits, a 1, and N'Z' are stored in the Status register (Figure 1006 page 207). The *ir 2,1,0* bits select the mcc report (Figure 1004). The mcc_report is the branch–do-not-branch decider for branch equations.

*NCZV status logic* (Figure 1005)
When all word bits of the alu or shift output equal zero, then Z=1.

When the alu carry out equals 1, then C=1.

When the alu or shifter output msb equals 1, then N=1. The assumption here is that the number is a signed number (it may not be ).

When inputs x and y msb both equal 1 (0), and the alu output msb equals 0 (1), then we have overflow V=1.

**Figure 1005 NCZV Logic**

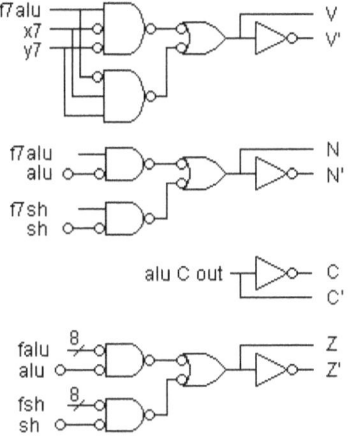

*TABLE 1004* CISC-8 Condition Code Equations. N C Z V are status bits
Equations produce mcc_reports (Figure 1003)

| mI sym | Result | status equation |
|--------|--------|-----------------|
| un | branch always | 1 |
| eq | result equal to zero | Z |
| ne | result not equal to zero | Z' |
| gt | result is >0 | N'Z' |
| lt | N result is negative, <0 | N |
| ge | result >0 or =0 | N' |
| cs | result sets carry | C |
| cc | result clears carry | C' |

# 10.4 User Instruction Formats

User instructions uI are represented by 8 bit words. *Fields* in any 8 bit user instruction word define uI opcodes, address modes, registers, and operations.

| 7 6 | 5 4 | 3 | 2 1 0 |
|---|---|---|---|
| 2 bits | 2 | 1 | 3 |
| Opcode | addr mode | reg | operation |

Opcode assignments

| <7:6> | uI type |
|---|---|
| 00 | LOAD |
| 01 | STORE |
| 10 | ALU/Shift |
| 11 | BR |

Address Modes

| <5:4> | mode |
|---|---|
| 00 | Inherent |
| 01 | Immediate |
| 10 | Direct |
| 11 | Index |

Register

| <3> | reg |
|---|---|
| 0 | $r_A$ |
| 1 | $r_X$ |

Operation field is 000 for Load and Store opcodes

Operation field for Shift, Alu, and Branch opcodes

| <2:0> | mode 0 | modes 1,2,3 | branch | br |
|---|---|---|---|---|
| 000 | SLL | ADD | BRun | 1 |
| 001 | SRL | SUB | BRcs | C |
| 010 | ROTL | INC | BRcc | C' |
| 011 | ROTR | DEC | BRlt | N |
| 100 | | NOT | BRge | N' |
| 101 | | AND | BReq | Z |
| 110 | | OR | BRne | Z' |
| 111 | | XOR | BRgt | N'Z' |

Examples

ADD *$r_X$(n) $r_A$     10110000

| 2 bits | 2 | 1 | 3 |
|---|---|---|---|
| Opcode | addr mode | reg | operation |
| 10 | 11 | 0 | 000 |

LOAD *$r_X$(n) $r_A$    00110000

| 2 bits | 2 | 1 | 3 |
|--------|------|-----|-----------|
| Opcode | addr mode | Reg | operation |
| 00 | 11 | 0 | 000 ignored |

STORE $r_A$ *$r_X$(n)    01110000

| 2 bits | 2 | 1 | 3 |
|--------|------|-----|-----------|
| Opcode | addr mode | Reg | operation |
| 01 | 11 | 0 | 000 ignore |

BReq *$r_X$(n)    11110101

| 2 bits | 2 | 1 | 3 |
|--------|------|-----|-----------|
| Opcode | addr mode | reg | operation |
| 11 | 11 | 0 | 101 |

SLL $r_A$    10000000

| 2 bits | 2 | 1 | 3 |
|--------|------|-----|-----------|
| Opcode | addr mode | reg | operation |
| 10 | 00 | 0 | 000 |

SUB *$r_X$(n) $r_A$    10110001

| 2 bits | 2 | 1 | 3 |
|--------|------|-----|-----------|
| Opcode | addr mode | reg | operation |
| 10 | 11 | 0 | 001 |

INC $r_A$    10100010

| 2 bits | 2 | 1 | 3 |
|--------|------|-----|-----------|
| Opcode | addr mode | reg | operation |
| 10 | 10 | 0 | 010 |

---

*Problem* 1004 Deduce the 8 bit word for the uI *LOAD 2B $r_A$*.

*Problem* 1005 Deduce the 8 bit word for the uI *STORE $r_A$ FF*.

*Problem* 1006 Deduce the 8 bit word for the uI *SUB A8 $r_A$*.

*Problem* 1007 Deduce the 8 bit word for the uI *BRne −5*.

*Problem* 1008 Deduce the 8 bit words for the uI *XOR $r_A$ *$r_X$(7)*.

---

## 10.5 Micro Instructions

User instructions uI and associated literals are fetched from memory, and executed by a sequence of *micro instructions mI* that are stored in the logic circuits of the computer control (Figure 1010 page 209).

> More complex computers store *mI lists* in a ROM, which is accessed as the uI is executed. This method requires a separate *micro data path* and a *micro data path controller*, which we do not design here. This *micro* is a hidden internal computer controlling the computer.

We refer to the sequence of mI as the *mI list* for the uI. *An mI list is required for each version of a uI.* E.g. the mI list for ADD is different for each address mode.

Each micro instruction *configures* the user data path, which implements mI task(s). For example: fetching a word representing a uI requires the pc to address the memory (mar←pc), and to add 1 to the *contents* of the pc (pc←pc+1) so that the pc addresses the next word in the program. Here is one way to represent this mI. *Each mI requires time to execute, which is the period of the mI clock that is derived from the cpu clock* (Figure 1007 page 208)

| mI clock period | actions |
|---|---|
| $n_0$ | $mar \leftarrow pc, \ pc \leftarrow pc+1$ |

> The actions are executed in parallel *at the clock edge* at the end of mI clock period $n_0$.

The user datapath uDP is configured to execute micro instructions by asserting various combinations of its control lines. The uDP control lines are configured by cpu control outputs (Figure 1010 page 209).

| mI clock period | actions | | asserted uDP inputs |
|---|---|---|---|
| $n_0$ | $mar \leftarrow pc$ | $pc \leftarrow pc+1$ | ldmar spcz incpc |

> All ← actions occur *on the clock edge **at the end** of mI clock period $n_K$.*

*Executing a uI* The *load iw reg* mI list (see list below) has 4 micro instructions mI executed in four *mI clock periods* $n_0$ to $n_3$ (10.7 page 208) The mI actions are as follows. (Address $m$ is in the pc.)

$n_0$ This mI stores the pc in the mar *on the clock edge at the end of clock period $n_0$* so that the pc address $m$ now in the mar addresses the uI word representing *load iw reg* in the memory at address m. In parallel the pc is incremented to $m+1$ at the end of mI clock period $n_0$.

$n_1$ Read $r$ is asserted during $n_1$. Read $r$ produces the *load iw reg* uI word at the memory q output before $n_1$ ends so that the *load iw reg* uI is loaded *on the clock edge at the end of period $n_1$* into the ir (Figure 1007 page 208).

$n_2$ A copy of the mI executed in $n_0$ is executed in $n_2$. This time the pc addresses the immediate word iw at address $m+1$ (Figure 1002 page 193).

$n_3$ The immediate word iw at the memory q output is stored in register reg ($r_A$ or $r_X$) *on the clock edge at the end of clock period $n_3$*. The mI clock counter $n$ is reset to zero at the end of mI clock period $n_3$ so that it is in state 0 at the start of the next mI list that will execute the next uI.

| $n$ | load uI *(immediate mode)* | *action* |
|---|---|---|
| $n_0$ | $mar \leftarrow pc, pc \leftarrow pc+1$ | *address the next uI in memory, inc the pc* |
| $n_1$ | $ir \leftarrow M(mar), r$ | *read mem, store the uI in the ir* |
| $n_2$ | $mar \leftarrow pc, pc \leftarrow pc+1$ | *address the immediate word, inc the pc* |
| $n_3$ | $reg \leftarrow M(mar), r, n \leftarrow 0$ | *read mem, store the iw in $r_A$ or $r_X$, clear mI counter to 0* |

> The memory read operation $r$ *must* produce the data by the end of any $n_k$. In other words the mI clock period is longer than a memory cycle.

*Problem* 1009 Write the mI list for the uI *XOR $r_A$ *$r_X$(7)*.

*Problem* 1010 Write the mI list for the uI *BRun *$r_X$(7)*.

## Load mI lists $\quad$ (reg is $r_A$ or $r_X$)

| $\underline{n}$ | load ul (inherent mode) | | |
|---|---|---|---|
| $n_0$ | not used | | |
| $n_1$ | – | | |
| $n_2$ | – | | |
| $n_3$ | – | | |

| $\underline{n}$ | load iw reg (immediate mode) | datapath commands | fetch iw to register |
|---|---|---|---|
| $n_0$ | $mar \leftarrow pc, \ pc \leftarrow pc+1$ | ldmar spcz incpc | ul address to mar |
| $n_1$ | $ir \leftarrow M(mar), \ r$ | ldir sqz r | ul to ir |
| $n_2$ | $mar \leftarrow pc, \ pc \leftarrow pc+1$ | ldmar spcz incpc | iw address to mar |
| $n_3$ | $reg \leftarrow M(mar), \ r, \ n \leftarrow 0$ | ldreg sqz r zeron | iw to register |

| $\underline{n}$ | load addr reg (direct mode) | datapath commands | fetch (addr) to register |
|---|---|---|---|
| $n_0$ | $mar \leftarrow pc, \ pc \leftarrow pc+1$ | ldmar spcz incpc | ul address to mar |
| $n_1$ | $ir \leftarrow M(mar), \ r$ | ldir sqz r | ul to ir |
| $n_2$ | $mar \leftarrow pc, \ pc \leftarrow pc+1$ | ldmar spcz incpc | addr address to mar |
| $n_3$ | $mar \leftarrow M(mar), \ r$ | ldmar sqz r | addr to mar |
| $n_4$ | $reg \leftarrow M(mar), \ r, \ n \leftarrow 0$ | ldreg sqz r zeron | (addr) to reg |

| $\underline{n}$ | load $*r_X(n)$ reg (index mode) | datapath commands | fetch $(*r_X(n))$ |
|---|---|---|---|
| $n_0$ | $mar \leftarrow pc, \ pc \leftarrow pc+1$ | ldmar spcz incpc | ul addr to mar |
| $n_1$ | $ir \leftarrow M(mar), \ r,$ | ldir sqz r | ul to ir |
| $n_2$ | $mar \leftarrow pc, \ pc \leftarrow pc+1$ | ldmar spcz incpc | n addr to mar |
| $n_3$ | $r_T \leftarrow M(mar), \ r$ (store $n$ in $r_T$) | ldrt sqz r | n to $r_T$ |
| $n_4$ | $mar \leftarrow r_X + r_T$ | ldmar saz add srxx srty | index addr to mar |
| $n_5$ | $reg \leftarrow M(mar), \ r, \ n \leftarrow 0$ | ldreg sqz r zeron | data to reg |

**Store mI lists**     (reg is $r_A$ or $r_X$)

| $\underline{n}$ | store uI  (inherent  mode) |
|---|---|
| $n_0$ | not used |
| $n_1$ | — |
| $n_2$ | — |
| $n_3$ | — |

| $\underline{n}$ | store uI  (immediate  mode) |
|---|---|
| $n_0$ | not used |
| $n_1$ | — |
| $n_2$ | — |
| $n_3$ | — |

| $\underline{n}$ | store reg addr (direct mode) | datapath commands |
|---|---|---|
| $n_0$ | $mar \leftarrow pc, \; pc \leftarrow pc+1$ | ldmar  spcz  incpc |
| $n_1$ | $ir \leftarrow M(mar), \; r$ | ldir  sqz  r |
| $n_2$ | $mar \leftarrow pc, \; pc \leftarrow pc+1$ | ldmar  spcz  incpc |
| $n_3$ | $mar \leftarrow M(mar), \; r$ | ldmar  sqz  r |
| $n_4$ | $M(mar) \leftarrow reg, w, \; n \leftarrow 0$ | sregx  sxz  w  zeron |

| $\underline{n}$ | store reg $* r_X(n)$ (index  mode) | datapath commands |
|---|---|---|
| $n_0$ | $mar \leftarrow pc, \; pc \leftarrow pc+1$ | ldmar  spcz  incpc |
| $n_1$ | $ir \leftarrow M(mar), \; r$ | ldir  sqz  r |
| $n_2$ | $mar \leftarrow pc, \; pc \leftarrow pc+1$ | ldmar  spcz  incpc |
| $n_3$ | $r_T \leftarrow M(mar), \; r$ (store n in $r_T$) | ldrt  sqz  r |
| $n_4$ | $mar \leftarrow r_X + r_T$ | ldmar  saz  add  srxx  srty |
| $n_5$ | $M(mar) \leftarrow reg, \; w, \; n \leftarrow 0$ | sregx  sxz  w  zeron |

## ALU mI lists   (reg is $r_A$ or $r_X$)

| $n$ | alu unary ops (inherent mode) | datapath commands |
|---|---|---|
| $n_0$ | $mar \leftarrow pc,\ pc \leftarrow pc+1$ | ldmar  spcz  incpc |
| $n_1$ | $ir \leftarrow M(mar),\ r$ | ldir  sqz  r |
| $n_2$ | $reg \leftarrow shop\ reg,\ n \leftarrow 0$ | ldreg  saz  shqp  sregx  zeron  ldst |

| $n$ | alu iw reg (immediate mode) | datapath commands |
|---|---|---|
| $n_0$ | $mar \leftarrow pc,\ pc \leftarrow pc+1$ | ldmar  spcz  incpc |
| $n_1$ | $ir \leftarrow M(mar),\ r$ | ldir  sqz  r |
| $n_2$ | $mar \leftarrow pc,\ pc \leftarrow pc+1$ | ldmar  spcz  incpc |
| $n_3$ | $r_T \leftarrow M(mar),\ r$ | ldrt  sqz  r |
| $n_4$ | $reg \leftarrow reg\ aluop\ r_T,\ n \leftarrow 0$ | ldreg  saz  aluop  srty  sregx  zeron  ldst |

| $n$ | alu addr reg (direct mode) | datapath commands |
|---|---|---|
| $n_0$ | $mar \leftarrow pc,\ pc \leftarrow pc+1$ | ldmar  spcz  incpc |
| $n_1$ | $ir \leftarrow M(mar),\ r$ | ldir  sqz  r |
| $n_2$ | $mar \leftarrow pc,\ pc \leftarrow pc+1$ | ldmar  spcz  incpc |
| $n_3$ | $mar \leftarrow M(mar),\ r$ | ldmar  sqz  r |
| $n_4$ | $r_T \leftarrow M(mar),\ r$ | ldrt  sqz  r |
| $n_5$ | $reg \leftarrow reg\ aluop\ r_T,\ n \leftarrow 0$ | ldreg  saz  aluop  srty  sregx  zeron  ldst |

| $n$ | alu $*r_X(n)$ reg (index mode) | datapath commands |
|---|---|---|
| $n_0$ | $mar \leftarrow pc,\ pc \leftarrow pc+1$ | ldmar  spcz  incpc |
| $n_1$ | $ir \leftarrow M(mar),\ r$ | ldir  sqz  r |
| $n_2$ | $mar \leftarrow pc,\ pc \leftarrow pc+1$ | ldmar  spcz  incpc |
| $n_3$ | $r_T \leftarrow M(mar),\ r$ | ldrt  sqz  r |
| $n_4$ | $mar \leftarrow r_X + r_T$ | ldmar  saz  add  srty  sxx |
| $n_5$ | $r_T \leftarrow M(mar),\ r$ | ldrt  sqz  r |
| $n_6$ | $reg \leftarrow reg\ aluop\ r_T,\ n \leftarrow 0$ | ldreg  saz  aluop  srty  sregx  zeron  ldst |

**Branch mI Lists**    Note: a prior uI sets the status bits.
*Note: mcc=mcc_report here*

| $\underline{n}$ | br uI *(inherent mode)* |
|---|---|
| $n_0$ | *not used* |
| $n_1$ | – |
| $n_2$ | – |
| $n_3$ | – |

| $\underline{n}$ | br iw *(immediate mode)* | *datapath commands* |
|---|---|---|
| $n_0$ | *mar* ← *pc, pc* ← *pc* +1 | *ldmar  spcz  incpc* |
| $n_1$ | *ir* ← *M*(*mar*), *r* | *ldir  sqz  r* |
| $n_2$ | *mar* ← *pc, pc* ← *pc* +1 | *ldmar  spcz  incpc* |
| $n_3$ | *n* ← 0, *if  mcc* = 1, *then*  *pc* ← *M*(*mar*), *r* | *zeron* (*mcc  AND  ldpc*) *sqz  r* |

| $\underline{n}$ | br addr *(direct mode)* | *datapath commands* |
|---|---|---|
| $n_0$ | *mar* ← *pc, pc* ← *pc* +1 | *ldmar  spcz incpc* |
| $n_1$ | *ir* ← *M*(*mar*), *r* | *ldir  sqz  r* |
| $n_2$ | *mar* ← *pc, pc* ← *pc* +1 | *ldmar  spcz incpc* |
| $n_3$ | *mar* ← *M*(*mar*), *r* | *ldmar  sqz  r* |
| $n_4$ | *n* ← 0, *if  mcc* = 1, *then*  *pc* ← *M*(*mar*), *r* | *zeron* (*mcc  AND  ldpc*) *sqz  r* |

| $\underline{n}$ | br *$*r_X$(n) *(index mode)* | *datapath commands* |
|---|---|---|
| $n_0$ | *mar* ← *pc, pc* ← *pc* +1 | *ldmar  spcz  incpc* |
| $n_1$ | *ir* ← *M*(*mar*), *r* | *ldir  sqz  r* |
| $n_2$ | *mar* ← *pc, pc* ← *pc* +1 | *ldmar  spcz  incpc* |
| $n_3$ | $r_T$ ← *M*(*mar*), *r (store n in $r_T$)* | *ldrt  sqz  r* |
| $n_4$ | *mar* ← $r_X$ + $r_T$ | *ldmar  saz  add  srty  srxx* |
| $n_5$ | *n* ← 0, *if  mcc* = 1, *then*  *pc* ← *M*(*mar*), *r* | *zeron* (*mcc  AND  ldpc*) *sqz  r* |

## 10.6 From uI to User Data Path uDP

The user data path is a set of building blocks interconnected by data buses. The operands of the uI user instructions specify some of the circuit blocks, the address modes require other blocks, and the uI register transfer statements specify the remainder. In other words Table 1001 page 191 essentially specifies the computer we intend to design. In what follows we make arbitrary yet practical decisions as we demonstrate that the user data path (Figure 1006), and the cpu control implement the user instruction set (Figures 1008, 1009, 1010 pages 208, 209).

***Bus*** Two sources and one destination require three 8 bit buses. We arbitrarily selected an 8 bit word size for our scaled down computer. Performance dictates that the 8 bits are available simultaneously when a word is read from any source. In hardware terms this means 8 wires, one wire per bit, are required to connect a source to a destination. This collection of 8 wires is called a bus. The bus provides parallel access to all word bits. Parallel access to two sources requires two 8 wire buses x, y to receive outputs from the blocks. One result requires one 8 wire bus z to deliver inputs to the blocks (Figure 1006).

The user data path shows the tri-state gates required to isolate the outputs. Tri-state gates prevent outputs connected to the same bus from shorting to each other. E.g. active *saz* connects only the ALU output to the z bus (Figure 1006).

***Status*** An operator uI executes, and sets the N C Z V status bits consistent with the results of the operation. Outputs from the alu and shifter are inputs to the NCZV logic, which produces the N C Z V bits. The load command *ldst* stores the N C Z V bits in the NCZV status register. In turn the status bits are inputs to the mcc logic, which implements the cc equations (Table 1004 page 197) that produce the mcc_reports.

***ALU, Shifter, & Registers*** implement uI arithmetic and logic operations on data and addresses, which are stored in 8 bit registers $r_A$, $r_X$, $r_T$, mar, RAM, pc, and ir.

***Memory*** A random access memory is required for user programs and their data (Chapter 8). Reading a memory fetches the contents of a register.

Reading requires an address to select the register storing the word. The address is stored in the memory address register (mar) in order to hold the address for the duration of a memory cycle. Writing a memory requires data in addition to the address. Input d and output q may be merged.

**Figure 1006 User Data Path uDP**

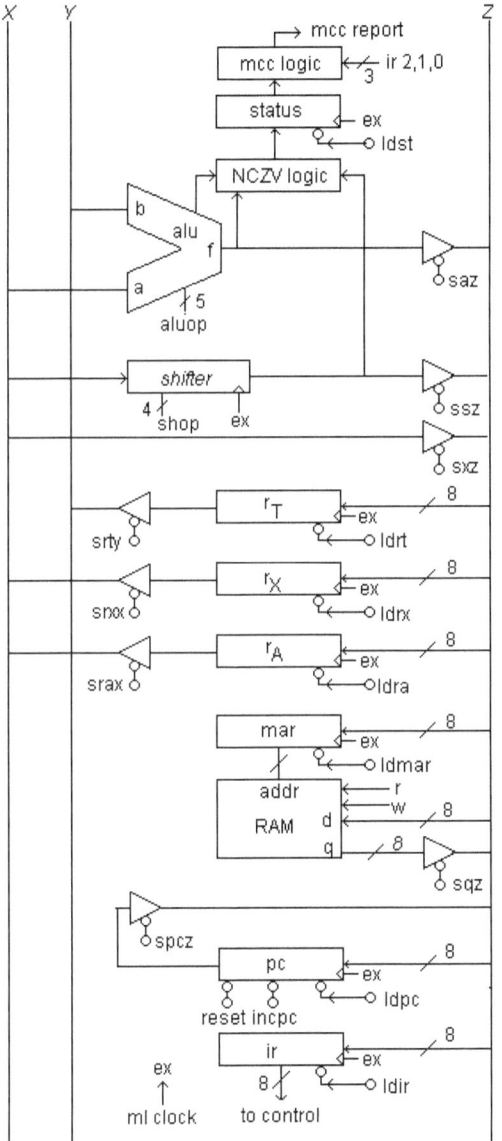

*Program Counter and Instruction Register ir:*
Every user instruction uI is fetched from memory, stored in the *ir* register, and executed. Fetching requires the address of every uI word in a program list. The easiest way to manage the list of uI is to *store* the uI words and their literals *sequentially* in memory.

A counting register, *the program counter, the pc,* is loaded with the address of the first uI word in the program.

The *pc* is incremented after each fetch of a uI or literal from the uI program list. In other words the *pc* always points to the next item in the program list.

Fetched uI are stored in the user instruction register (ir). Literals are stored in *mar, $r_A$, $r_X$, $r_T$.*

# 10.7 CISC Control

The mI read the memory to fetch user instructions and perhaps an immediate iw word or index number. Consequently the time a mI requires is the same as a memory cycle with several wait states (Section 8.1 page 155). The control uses *inc n* (Figure 1008) to increment the $n_K$ number at the end of every mI clock period so that the system knows which mI in the list is executing. The last mI in a mI list signals it is the last mI by issuing the *zeron* signal (Figures 1007, 1008). When *zeron=T* conditional output *n=0* resets n to zero so that if *pend=F* the next mI list starts executing the first mI in its list.

**Figure 1007 mI Timing**

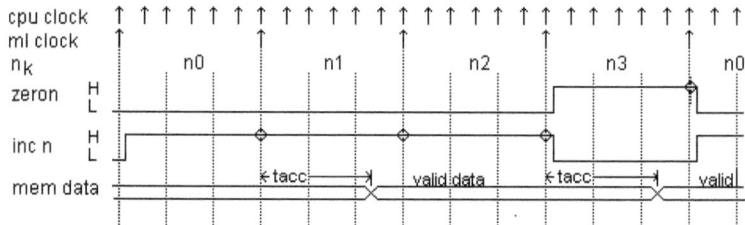

**Figure 1008 Control ASM**

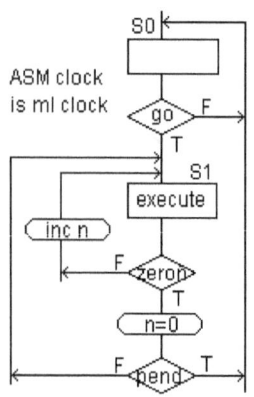

An elementary ASM, *clocked by the mI clock*, implements control of the mI timing (Figure 1008). Here we are only concerned with user datapath control. The mI clock executes the functions setup in the datapath (Figure 1006) by the cpu control (Figure 1010).

The last mI in any mI list clears the timer to zero so that execution of each uI starts with the first mI in the mI list in time period $n_0$. Consequently each uI only uses a number of clock periods equal to the number of mI in the list.

The mI clock is derived from the cpu clock by dividing by 6 to have a time period equal to the time required by a memory cycle with several wait states (Figures 1007, 1009). A mI list line counter and a 3 to 8 decoder produce the $n_0$ to $n_6$ required to select a mI in a mI list. Asserted $n_K$ activate terms in cpu control equations. The inputs and outputs to cpu control and datapath are shown in Figure 1009.

**Figure 1009 Control and User Datapath**

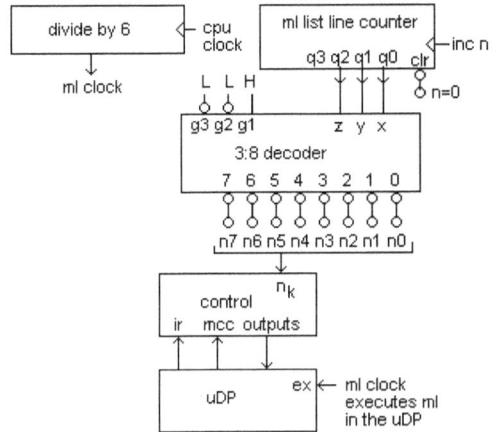

Branch uI *BRun halt* in effect activates *pend* (Figure 1008) to signal the end of the program and a return to the operating system.

**Figure 1010 CPU Control**

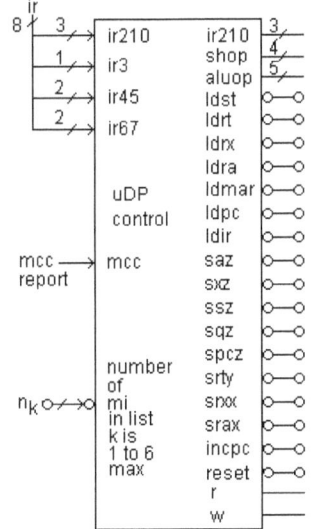

*The mI lists are stored in the control* in the form of Boolean equations (Figure 1010). The uI bits in the ir configure the cpu control, which decodes the ir bits (Figure 1011). Decoding asserts variables in logic equations. Each equation is a cpu control output to the datapath. For example the *w* equation controls memory writes by the fourth and fifth mI in STORE mI lists using the direct and index address modes.

$$w = n_4[store \cdot dir] + n_5[store \cdot indx]$$

The cpu control outputs are inputs to the user datapath, which is configured by those inputs (Figure 1009).

The uI bits in the ir and the mcc reports are sent from the uDP to the control (Figures 1009, 1010).

## Figure 1011 CPU Control IR Decode

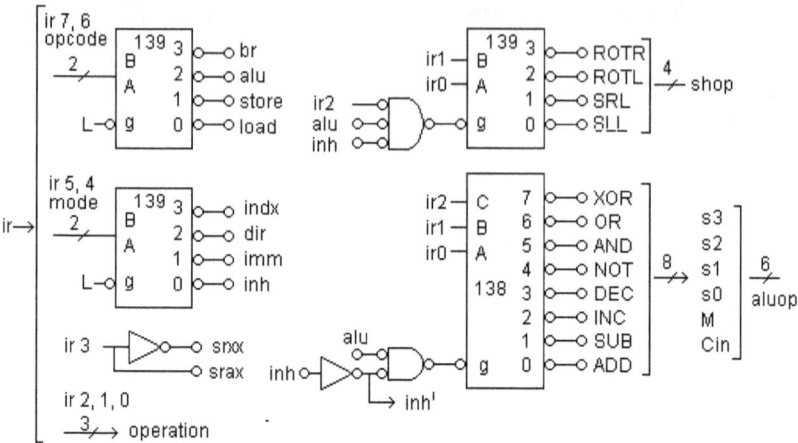

Decoded ir bits (Figure 1011) assert variables representing the current uI. The aluop equations use asserted variables to program the alu. The *standard circuit 181* ALU opcodes $S_{3210}$ are listed in the table.

| aluop | $S_{3210}$ | M | $C_{in}$ |
|-------|-----------|---|----------|
| NOT | LLLL | H | – |
| AND | HLHH | H | – |
| OR | HHHL | H | – |
| XOR | LHHL | H | – |
| ADD | HLLH | L | H |
| SUB | LHHL | L | L |
| INC | LLLL | L | L |
| DEC | HHHH | L | H |
| ZERO | LLHH | L | L |

Aluop equations.

$$s_3 = AND + OR + ADD + DEC$$
$$s_2 = OR + XOR + SUB + DEC$$
$$s_1 = AND + OR + XOR + ZERO + SUB + DEC$$
$$s_0 = AND + ZERO + ADD + DEC$$
$$M = NOT + AND + OR + XOR$$
$$C_{in} = ADD + DEC$$

The 13 mI lists (10.5 pages 202-205) specify actions implementing them so that the mI actions configure the user data path uDP. These actions are included in the cpu control equations, which configure the uDP. For example here are equations for ldir, ldpc, and incpc.

**ldir** uDP input *ldir* is asserted whenever the action *ir← M(mar)* occurs. We can derive the logic equation for *ldir* by studying the 13 mI lists.

The 13 mI lists show that the *ir* is loaded only in clock period $n_1$. The *ir* is loaded in the Load mI or Branch mI immediate, direct, and index modes.

The *ir* is also loaded in the Store mI direct and index modes, and in all Alu modes.

$$ldir = n_1[load(imm + dir + indx) + store(dir + indx) + alu + br(imm + dir + indx)]$$

**ldpc** The pc is loaded with a new address only when a branch is taken. The action is $pc \leftarrow M(mar)$.

$$ldpc = n_3[br(imm \cdot mcc \_ report)] + n_4[br(dir \cdot mcc \_ report)] + n_5[br(indx \cdot mcc \_ report)]$$

**incpc** The action $pc \leftarrow pc+1$ adds 1 to the address in the pc in mI clock periods $n_0$ and $n_2$.

$$incpc = n_0[load(imm + dir + indx) + store(dir + indx) + alu \quad + br(imm + dir + indx)]$$
$$+ n_2[load(imm + dir + indx) + store(dir + indx) + (alu + br)(imm + dir + indx)]$$

---

Note: do not simplify the logic equations.
*Problem* 1011 Write the logic equation for user data path input *srty*.

*Problem* 1012 Write the logic equation for user data path input *sqz*.

*Problem* 1013 Write the logic equation for user data path input *spcz*.

*Problem* 1014 Write the logic equation for user data path input *zeron*.

*Problem* 1015 Write the logic equation for user data path input read *r*.

*Problem* 1016 Write the logic equation for user data path input *ldst*.

*Problem* 1017 Write the logic equation for user data path input *ldmar*.

*Problem* 1018 Write the logic equation for user data path input *ldra*.

*Problem* 1019 Write the logic equation for user data path input *saz*.

---

# 10.8 Machine Language Programs

The following program is written in the *assembly language* of our CISC-8 computer. The uI list written in assembly language is translated into *machine language*. The CISC-8 *assembler program* converts the uI list to a list of binary words we store in the memory.

*Sample program : Clear memory to 0 from address $E0_{16}$ to $EF_{16}$*

| label | uI | action | |
|-------|-----|--------|---|
| | LOAD $E0_{16}$ $r_X$ | $r_X \leftarrow E0$ | (*E0 now in* $r_X$) |
| clear | LOAD 0 $r_A$ | $r_A \leftarrow 0$ | |
| | STORE $r_A$ * $r_X$ (0) | $M(r_X) \leftarrow r_A$ (0 *stored in addr* E0_plus_inc) | |
| | INC $r_X$ | $r_X \leftarrow r_X +1$ | (*$r_X$ ranges from* E0 *to* F0) |
| | LOAD $F0_{16}$ $r_A$ | $r_A \leftarrow F0 = EF +1$ | |
| | SUB $r_X$ $r_A$ | $r_A \leftarrow r_A - r_X$ (*don't care about* $r_A$ *value*) | |
| | BRne clear | go to clear to repeat cycle if $r_X \neq F0$ | |
| | BRun halt | end program | |

*Translation to binary examples*

LOAD EF $r_A$

| Opcode | addr mode | reg | operation |
|--------|-----------|-----|-----------|
| Load | immediate | rA | n/a |
| 00 | 01 | 0 | 000 |

$uI = 00010000_2 = 10_{16}$

$iw = 11101111_2 = EF_{16}$

STORE $r_A$ *$r_X$(0)

| Opcode | addr mode | reg | operation |
|--------|-----------|-----|-----------|
| Store | index | rA | n/a |
| 01 | 11 | 0 | 000 |

$uI = 01110000_2 = 70_{16}$

$n = 00000000_2 = 00_{16}$

INC $r_X$

| Opcode | addr mode | reg | Operation |
|--------|-----------|-----|-----------|
| Alu | inherent | rX | INC |
| 10 | 00 | 1 | 010 |

$uI = 10001010_2 = 8A_{16}$

# 11 Verilog - a Language for Digital Design[1]

Now that gate level/schematic design of some elementary and some complex algorithms has been presented in Chapters 1 to 10 we turn to the major design issue of the day. *How do we design a digital system when the implementation requires a very, very large number of gates?*

For large circuit designs *text capture* of digital designs is preferred to *schematic capture*. The Verilog language syntax offers clarity and savings in design time. The design process is enhanced by the integration of Verilog simulation, and synthesis tools which implement many of the design details, which we do not discuss.

The purpose of a hardware description language (HDL) such as Verilog is to describe, specify, the function of hardware independently of the implementation. The great step forward with an HDL was the recognition that a single language could be used to describe the function of the design, simulate the design, and be synthesized into the implementation.

In turn a compiler operating on a Verilog program produces file(s) that describe integrated circuit masks for a very large scale integrated chip (VLSI). Or produces files that describe a digital circuit of flip-flops and gates.

In turn a synthesis engine converts Verilog program into gate level netlists with timing information that describe the gate level netlist for a very large scale integrated chip VLSI, PLA, or FPGA. The gate level netlist is run through a place and route tool which creates the layout design for the target device.

The combination of Verilog and a compiler is a complete design system with essentially no limit on digital circuit complexity.

Verilog programs are based on the hierarchy of a design. The Verilog structures which build the hierarchy are modules with ports. The ports describe how the modules are interconnected.

What follows is a brief introduction to Verilog.

---

[1] We assume the reader knows how to write computer programs.

# 11.1 Module Hierarchy

Verilog programs are based on the hierarchy of a design. The Verilog structures which build the hierarchy are *modules with ports*.

A Verilog program represents a digital system as a group of modules. The *main*, or *top*, module instantiates defined modules that represent the digital circuit to be designed. Each module has input and output ports. The *interface* to other modules shows how the ports are interconnected. In this way a tree of modules is constructed (Figure 1101).

A module which is composed of other module instances is called a parent module, and the instances are called child modules.

For example, consider the modules main, ckt1, ckt2, and subckt1 where main instantiates ckt1 and ckt2, and ckt2 instantiates subckt1 (Figure 1101).

**Figure 1101 Module Hierarchy**

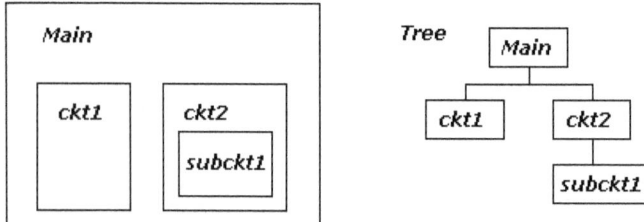

When module Main instantiates module ckt1, module Main creates an instance of module ckt1 as a child module in module Main.

> A module is written at one of four levels of Abstraction: Behavioral/Algorithmic, Dataflow, Gate, Switch

*Important* the contents of a module, the internals, can be modified without affecting the rest of the design.

> The examples do not necessarily represent best programming practices.

# 11.2 Modules

Modules are specified behaviorally, structurally, or in combination. A *behavioral specification* defines the behavior of a module with programming language elements such as if-then, case, and so forth. A *structural specification* expresses the behavior of a module as a hardware hierarchy of gates and with continuous assign statements.

Modules represent blocks of hardware ranging from gates to complete systems such as a computer. A module is defined as follows.

*Modules-top level* There is one exception to the rule that *the module definition does not create/instantiate a module.* The exception is for top level modules. A top level module is not instantiated by any other module or itself. The top level module name is also its instance name.

*Modules-instances* The original module SRlatch is a template, which does not create a module. In module top, SRlatch instantiates latch1(qout, zoutbar, setbar, resetbar); (Example 1101 page 216)

```
module <parameters> <module name> (<port list>);
    <declares>
    <module internals>
endmodule
```

<parameters> are values of parameters passed to the instance.

<module name> is an identifier that uniquely names the instantiated module.

<port list> is a list of input, output, and (bi-directional) inout ports which are used to define connections to other modules.

<declares> specifies data objects such as wires, registers, memories, and function constructs.

<module internals> can include *initial* constructs, *always* constructs, continuous assignments, and instances of modules.

Once a module is instantiated in a program, the module exists for the life of the program.

Digital Design

---

*Example 1101* (Figure 501) Study examples to learn Verilog.

```
module top;
    wire qout, zoutbar;                // bar to indicate active low
    wire setbar, resetbar;
    SRlatch latch1(qout, zoutbar, setbar, resetbar);
                    //instantiated names can be different
endmodule

module SRlatch(q, zbar, sbar, rbar);
    output q, zbar;                    //latch outputs are wires
    input sbar, rbar;                  //latch inputs are wires
    nand n1(q, sbar, zbar);            //nand is a Verilog primitive
    nand n2(zbar, q, rbar);
endmodule
```

## Figure 501 SR Latch

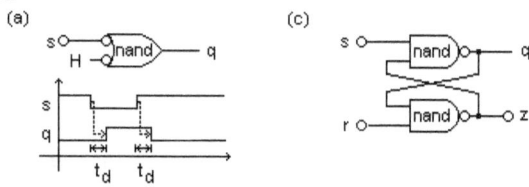

---

*Example 1102* (Figure 1102)
```
module top;
    wire fout, zwire;
    wire a1, a0;           //wire or reg, circuit dependent
    Mux2_1 mux1(fout, zwire, a1, a0);
endmodule
```

```
module Mux2_1(f, z, a1, a0);
    output f,;
    input z, a1, a0;
    wire zbar, x, y;            //internal
    not inv(zbar, z);
    nand n1(x, z, a1);
    nand n2(y, zbar, a0);
    nand n3(f, x, y);
endmodule
```

**Figure 1102  2:1 Mux**

---

216

## *Verilog Operators*

| Operator | Symbol | Operation | Operands |
|---|---|---|---|
| Arithmetic | * | multiply | two |
| | / | divide | two |
| | + | add | two |
| | − | subtract | two |
| | % | modulus | two |
| Bitwise | ~ | bitwise negation | one |
| | & | bitwise and | two |
| | \| | bitwise or | two |
| | ^ | bitwise xor | two |
| | ^~ or ~^ | bitwise xnor | two |
| Concatenation | { } | concatenation | two or more |
| Conditional | ?: | conditional | three |
| Equality | == | equality | two |
| | != | inequality | two |
| | === | case equality | two |
| | !== | case inequality | two |
| Logical | ! | logical negation | one |
| | && | logical and | two |
| | \|\| | logical or | two |
| Relational | > | greater than | two |
| | < | less than | two |
| | >= | greater than or = | two |
| | <= | less than or = | two |
| Reduction | & | reduction and | one |
| | ~& | reduction nand | one |
| | \| | reduction or | one |
| | ~\| | reduction nor | one |
| | ^ | reduction xor | one |
| | ^~ or ~^ | reduction xnor | one |
| Replication | { { } } | replication | two or more |
| Shift | >> | right shift | two |
| | << | left shift | two |

# 11.3 Module Ports

Ports are Verilog structures that pass data between modules. Port lists can be interpreted as wires connecting modules. The types of port connections are *input, output, or inout* (bi-directional).

*Port list* Ports are named in the port list attached to the module name. Port direction is declared by type: input, output, inout. For example define *decoder2:4*.

module decoder2_4 (fout,y,x,g);
output [3:0] fout;
input x, y, g;
assign fout[0] = ~(~y & ~x & ~g),
fout[1] = ~(~y & x & ~g),
fout[2] = ~(y & ~x & ~g),
fout[3] = ~(y & x & ~g);
endmodule

**Figure 1105 Decoder**

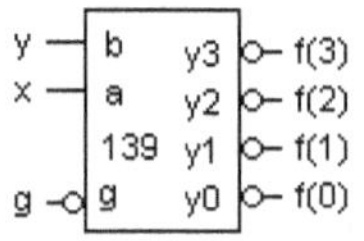

Verilog imposes no restrictions on the order of ports in a port list in the definition. That is, inputs, outputs, and inouts can be mixed in any order in the definition. *In the instantiation, however, the order must match the definition.* Any port can be omitted in the instantiation by double commas.

*Warning* - do not omit a port in an original module definition.

*Comments on module decoder2_4* Output *fout* is declared as a 4 bit wire, which is a 4 wire bus. The inputs *x, y, g* are wires. The ~ is the bitwise NOT function, and *~y* means *y* is active low. The & is the bitwise AND function. Active low *~g* enables the decoder. Think of *x* and *y* as address lines with binary values 00, 01, 10, 11, so that *~yx* = 01.

The *continuous assignment* key word assign defines a relation between the left hand side *variable* and the *expression* on the right-hand side. Continuous assignments are always active. Observe how commas are used.

*Port assignments - Errors* The most common error is omitting a port assignment. Note that since double comma omissions are legal, the simulator does not report them as an error. This is a double comma omission of y (fout, ,x,g) Careful!

# 11.4 Gates

Verilog includes logic gates as *predefined primitives*, which do not need a defining module. The primitives can be instantiated as if they are defined modules.

*-And/Or* gates have one output and 2 to n inputs (Figure 1103). A gate output is evaluated immediately when an input changes.
*-Buf/Not* gates have 1 or more outputs and 1 input (Figure 1104). A gate output is evaluated immediately when an input changes.

**Figure 1103 And/Or Gates**          **Figure 1104 Buf/Not Gates**

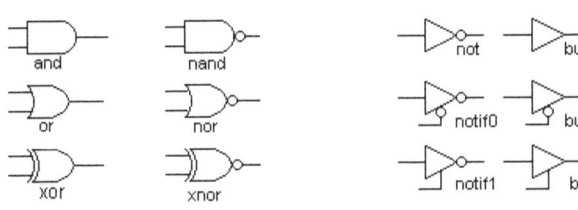

---

*Example 1103* (Figure 408)
```
module Mux4_1(f, y, z, a3, a2, a1, a0);
    output f;
    input y, z, a3, a2, a1, a0;
    not inv1(zbar, z);
    not inv2(ybar, y);
    wire ybar, zbar;          //internal wires
    nand n1(q0, ybar, zbar, a0);
    nand n2(q1, ybar, z, a1);
    nand n4(q2, y, zbar, a2);
    nand n5(q3, y, z, a3);
    wire q3, q2, q1, q0;    //internal wires
    nand n3(f, q3, q2, q1, q0);
endmodule
```

**Figure 408 Mux 4:1**

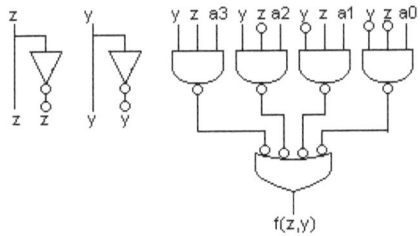

---

# 11.5 Data Flow

You can create a circuit design based on the flow of data from register to register. The advantage is that *synthesis* tools can create a gate level circuit from a data flow description. Statements such as this one are not obvious.

*Continuous Assignments* The continuous assignment key word assign defines a relation between the left hand side *variable* and the *expression* on the right-hand side. (Continuous assignments are always active.)

*Assign syntax* (drive_strength) (delay) net_assignments;

Examples:
assign z = x & y;
assign (strong1, pull0) (#2) net3 = net0;
assign (#1) d3 = q3 ^ (q2 & q1 & q0);

When continuous assignments change the state of registers, they implement *sequential logic*.

*Net on Left* In continuous assignments the left-hand side must be a net (wire, vector). The reason for this is that changes to a net can happen asynchronously. Any time anything on the right-hand side changes, the left-hand side changes its value. This is why registers are not allowed on the left side, because registers get new values only at clock edges.

During simulation, when any component of the right-hand side expression changes in value, the right-hand side expression is re-evaluated, and the value is assigned to the left-hand side net (e.g. wire). The assignment may be delayed by a specified delay amount, and it may have a given strength. If no delay is specified, the assignment happens at the current simulation time, and if no strength is specified, the default value is used. In other words assign is used to model combinatorial circuits whose present output follows any input changes.

*drive_strength* Scalar wires can have different strength levels. A strength level is associated with each driver on the net. If a strength is not specified, then the default for that driver is used. The default is usually strong.

*Delays* A delay can be assigned to a continuous assignment. Delay statements use the # character. The examples below show that the variable on the left changes 5 time units *after* the right hand expression changes.

•Regular form:
assign (#5) d3 = q3 ^ (q2 & q1 & q0);   // ^ is xor

•Implicit form:
wire d3;
assign (#5) d3 = q3 ^ (q2 & q1 & q0);

•Net delay form:
wire #5 d3;
assign d3 = q3 ^ (q2 & q1 & q0);

---

*Example 1104* (Compare to Example 1103, Figure 408)
module Mux4_1(f, y, z, a3, a2, a1, a0);
    output f;
    input y, z, a3, a2, a1, a0;

assign f = (~y & ~z & a0)|( ~y & z & a1)|
          ( y & ~z & a2)|( y & z & a3);
Endmodule
                    // the | is the OR function

**Figure 408 Mux 4:1**

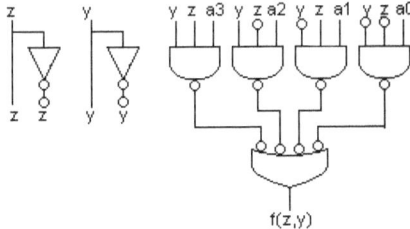

---

# 11.6 Behavioral

A state machine implements an algorithm. The algorithmic state machine (ASM) chart is a preferred way to represent the algorithm (Chapter 5). Verilog behavioral statements implement algorithmic state machines.

*Procedures* Verilog includes two basic procedure statements: initial and always. All other behavioral statements appear only inside initial and always statements.

*Initial* Initial blocks execute the initial statement starting at time 0. When the initial statement finishes execution, the initial block terminates. If there are several initial blocks, then each starts at time 0 so that they execute in parallel. Each block completes execution independently of the other initial blocks.

---

*Example 1105* Initial blocks
Blocking assignments ( = ) execute sequentially in time.
Delays add, because the next assignment in a list, such as c, starts *after* the prior assignment in the list b completes execution. Numbers are represented as
<no of bits>'(tick)<base><number> such as 4'b1101
// base can be 2,8,10,16 as binary, octal, decimal, hex
module Number;
    reg a, b, c, x, y, z;
    initial
        begin
                a = 1'b0;          // 1 bit binary a=0 at time 0
            #2 b = 1'b1;          // b=1 at time 2
            #6 c = 3'b111;       // c=7 at time 8
        end

    initial
        begin
                x = 16'd5;        // 16 bit decimal x=5 at time 0
            #5 y = 16'd6;        // y=6 at time 5
            #9 z = 16'd7;        // z=7 at time 14
        end
endmodule

---------------------------------------

| Time | statements executed | |
|------|------|------|
| 0 | a = 1'b0 | x = 16'd5 |
| 2 | b = 1'b1 | |
| 5 | | y = 16'd6 |
| 8 | c = 3'b111 | |
| 14 | | z = 16'd7 |

---

*Example 1106* Initial blocks
Blocking assignments ( = ) execute sequentially in time.
Delays add.

module Number;
reg a, b, c;

initial
    begin
        a = 0; b = 1; c = 2;        // notice the semicolons
    end
endmodule

-------------------------------------

| Time | statements executed |
|------|---------------------|
| 0    | a = 0               |
| 1    | b = 1               |
| 2    | c = 2               |

*Example 1107* Initial blocks
Non Blocking assignments ( <= ) start at time 0, and execute in parallel time.
Delays do not add.
Note: Blocking and non blocking here.

module Number;
reg a, b, c, x, y, z;

initial
    begin
        a <= 0;                // a=0 at time 0
        #2 b = 1; c <= 2;      // b=1 at time 2, c=2 at time 3
    end

initial
    begin
        x <= 4'b0101;          // x=5 at time 0
        #5 y <= 4'b0110;       // y=6 at time 5
        #9 z <= 4'b0111;       // z=7 at time 9, not 14
    end

endmodule

-------------------------------------

| Time | statements executed | |
|------|---------------------|---|
| 0    | a = 0               | x = 4'b0101 |
| 2    | b = 1  // default 32 bit decimal numbers | |
| 3    | c = 2               | |
| 5    |                     | y = 4'b0110 |
| 9    |                     | z = 4'b0111 |

223

Digital Design

*Always* always blocks start execution at time t=0. When the always block finishes execution, it starts executing again (a loop).

There may be any number of initial and always blocks in a module. All blocks start to execute at time 0. There is no guarantee that any statement will execute before or after any other statement that is not in the same block unless there is a time or event control to establish that relationship. The timing of modules is asynchronous. Enter *event control*.

*Event Control* @ The syntax @(event list ...) executes a *statement* or a *begin-end* block only after one or more events in the list occur. An event is a change in a variable. The change is a signal's positive edge, negative edge, or just a level change. Multiple events are listed by the **or** keyword. Event syntax is *posedge var_name*, *negedge var_name*, or just *var_name*.

---

*Syntax 1*
always @(event_1 or event_2 or ...)
    //wait for event_1 or event_2 or ... to execute
    begin
    ... *statements* ...
    end

*Syntax 2*
always @(event_1 or event_2 or ...)
    begin: name_for_block
    ... statements ...
    end

*Example 1108 Always blocks*
always @(x or y) //*if x or y change levels execute x or y*
    begin ... end
always @(posedge clk or negedge clear); /**edge-triggered on positive edge execute clock or negative edge execute clear.*/
    begin ... end

always begin       //think of this as *Syntax 3*
    @(posedge clock) state1 = nextstate1;
    @(posedge clock) state2 = nextstate2;
end

---

# 11.7 Four Bit Up Counter

A state machine has a next state combinatorial logic function, a state memory function, and an output combinatorial logic function (Figure 501).

**Figure 501 State Machine Block Diagram**

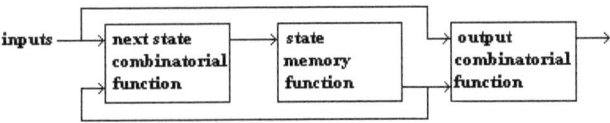

Studying formal language documents[1] is not productive for us, because how-to-write-code is *not* an explicit part of such documents. We prefer to jump into the water converting Algorithmic State Machine (ASM) designs to Verilog code, learning syntax as we proceeded. Consequently we learned a great deal about the whys and wherefores of writing Verilog code.

The four bit up counter next state design equations (7.2.3 pages 128, 131) for D flip-flops are

$$d_3 = c'gN_3 + c'g'(q_3 \, xor \, pq_2q_1q_0) \quad count \, up$$
$$d_2 = c'gN_2 + c'g'(q_2 \, xor \, pq_1q_0)$$
$$d_1 = c'gN_1 + c'g'(q_1 \, xor \, pq_0)$$
$$d_0 = c'gN_0 + c'g'(q_0 \, xor \, p)$$

where the present state is $q_3q_2q_1q_0$. Equation inputs are clock, c clear, g load, p count, and the four $N_k$ bits are data to load. The state number is stored in output register *output1*.

---

[1] IEEE Std 1364-2001 (Revision of IEEE Std 1364-1995)
IEEE Standard Verilog® Hardware Description Language
Note - This document is available as a pdf file, is 791 pages long, and does not have an index!

Digital Design

1) The Verilog model for the counter is represented by this module definition that is specified by the module syntax (11.2 page 215)

```
module ctr_up_4bit(clock, clear, load, count, data, output1);
    input clock, clear, load, count;
    input   [3:0] data;
    output [3:0] output1;
    ...(statements here)...
endmodule
```

2) Registers are data storage elements. They are variables that retain value. They are not equivalent to hardware registers.

Every state machine has a state, a next state, and state variables $q_k$. The items are registered so that their values are stored.

```
reg [3:0] state, next_state;
reg q3, q2, q1, q0; // four 1 bit reg
```

3) The three ASM functions (Figure 501 p225) are next_state_logic, state (the state memory), and output_logic (which in this example is the port output1).

next_state_logic is implemented as an always block (11.6 page 222) so that it can calculate the counter's next_state.

**If** clear is true the *next_state* is 0000
**Else-if** the count is incremented by one
**Else** the counter is *loaded* with the state number stored in *data*

The non-blocking assignment <= allows the four $q_k$ bits to be evaluated simultaneously.

4) always

Register *state* (the state_memory) is updated by *next_state*, which was calculated by next_state_logic.

Register output1 is updated by *state*.

Putting all steps together we get the following program.

```
module ctr_up_4bit(clock, clear, load, count, data, output1)

input clock, clear, load, count;
input [3:0] data;
output [3:0] output1;

reg [3:0] state, next_state;    // 4 bit reg
reg q3, q2, q1, q0; // four 1 bit reg

next_state = {q3, q2, q1, q0};    //concatenate 4 variables

always @ (clear or count or load)
  begin: next_state_logic //Name of this always procedure.
    if (clear) begin
      next_state <= 4'b0000;
    end
    else if (count) begin // eval bits simultaneously
      q3 <= q3 ^ (q2 & q1 & q0);        // ^ is xor
      q2 <= q2 ^ (q1 & q0);
      q1 <= q1 ^ q0;
      q0 <= q0 ^ 1;
      end
    else (load) begin
      next_state <= data;
      end
  end

always @(posedge clock)
  state = next_state; //one statement so can omit begin-end

always @(state)          //one statement, begin-end not omitted
  begin: output_logic
    output1 = state;
  end
endmodule
```

# 11.8 Shift Registers

Shift registers use the operators << and >> in conjunction with the nonblocking <= assignment operator. The size of the shift register is conveniently set by changing the value of one parameter number such as width. Therefore each instance of a shift register module can have a different width. *This is an example of the use of a constant that simplifies implementation of number changes.*

```
module shift(shiftOut, dataIn, shiftBy);
   parameter width = 4;
   output [width-1:0] shiftOut;
   input [width-1:0] dataIn, shiftBy;
   assign shiftOut <= dataIn << shiftBy;
endmodule
```

```
module rotl(q, clock, reset);
   parameter width = 4;
   output [width-1:0] q;       //width-1=4-1=3
   input clock, reset;
   reg [width-1:0] qr;
   always @(posedge clock or posedge reset)
      begin
        if (reset) qr <= width'b0;
        else begin
           qr <= qr << 1; // Left shift 1 position
           qr[0] <= qr[width-1];
           q <= qr;
           end
      end
endmodule
//Nonblocking means the old qr[width-1] is sent to qr[0].
```

```
module lfsr(q, clock, reset);
   output [width-1:0] q;
   input clock, reset;
   reg [width-1:0] qr;
   always @(posedge clk or posedge reset)
   begin
      if (reset) qr <= 0;
      else
      qr <= {qr[width-2:1], qr[width-1]^qr[width-2]}
/* The {brackets} are concatenation operators {} that form the new qr from elements of the
old qr. */
      q <= qr;
   end
endmodule
```

# 11.9 Two Bit Up Counter

1) The Verilog model for the counter is represented by module fig502asm.

module fig502asm (clock, c, p, Ready, Three, y, q);
    ….(port lists) …
    ….(statements here). …
endmodule

**Figure 502 ASM**

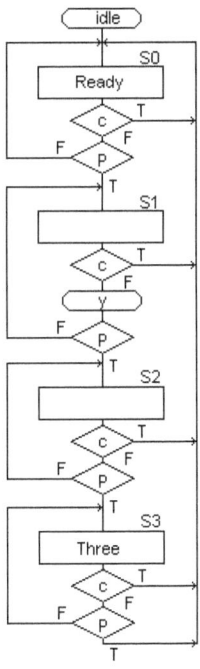

2) The module fig502asm has 3 1-bit input ports: clock, c (clear), p (count), 3 1-bit output ports Ready, Three, y, and one 4-bit output port q.

Variables are registered, because their stored contents are always available. Adding port state variables, and parameter declarations we get

module fig502asm (clock, c, p, Ready,
 Three, y, q);
input clock, c, p;
output Ready, Three, y, q;
reg [3:0] next_state, state, q;
parameter state0=0, state1=1, state2=2, state3=3;
… statements …
endmodule

3) Three ASM functions named next_state_logic, state_memory, output_logic are implemented as always blocks.

next_state_logic A case statement selects the present state that calculates the next state per the input variables $c$ and $p$.

state_memory Reg *state* is updated by *next_st* that was calculated by next_state_logic.

output_logic Output $q$ is updated by *state*. If-else statements calculate values for Ready, Three, and y

```
module fig602asm (clock, c, p, Ready, Three, y, q);

input clock, c, p;
output Ready, Three, y, q;
reg [3:0] next_st, state, q;
parameter state0=0, state1=1, state2=2, state3=3;

always @ (state or c or p)
    begin: next_state_logic
            case (state)
            state0: begin
                Ready=1;
                if (c) begin
                    next_st = state0;
                    end
                else if (p) begin
                    next_st = state1;
                    end
                    else begin
                        next_st = state0;
                    end
            end
            state1: begin
                if (c) begin
                    next_st = state0;
                    end
                else begin
                    y=1;
                    if (p) begin
                        next_st = state2;
                        end
                        else begin
                            next_st = state1;
                        end
                end
            end
            state2: begin
                if (c) begin
                    next_st = state0;
                    end
                else begin
                    if (p) begin
                        next_st = state3;
                        end
                        else begin
                            next_st = state2;
                        end
                end
            end
```

```
            state3 begin
                Three=1;
                if (c) begin
                    next_st = state0;
                    end
                else begin
                    if (p) begin
                        next_st = state0;
                        end
                    else begin
                        next_st = state3;
                        end
                    end
            default: begin
                    next_st = state0;
                    end
            endcase // default is optional but use it
        end

always @(posedge clock)
    begin: state_memory
    state = next_st;
    end

always @(state)
    begin: output_logic
            q = state;
            if (state==0) Ready=1 else Ready =0;
            if (state==1 & ~c) y=1 else y=0;
            if (state==3) Three=1 else Three=0;
    end

endmodule
```

# Answers to most of the problems

**1** What is the logic equation that keeps the ASM in State 3?

$S_3 \times kbd'$

**2** What is the logic equation that produces output N?

$S_0 \times R{<}N \times R{=}0$

**3** Write the truth table for *and*.

**4** Write the truth table for *not*.

**5** Write the truth table for the *or*.

| Logical | | | AND | OR | NOT | | Physical | | | AND | OR | NOT |
|---|---|---|---|---|---|---|---|---|---|---|---|---|
| Row | x | y | xy | x + y | y' | | Row | x | y | xy | x + y | y' |
| 0 | 0 | 0 | 0 | 0 | 1 | | 0 | L | L | L | L | H |
| 1 | 0 | 1 | 0 | 1 | 0 | | 1 | L | H | L | H | L |
| 2 | 1 | 0 | 0 | 1 | | | 2 | H | L | L | H | |
| 3 | 1 | 1 | 1 | 1 | | | 3 | H | H | H | H | |

**6** Find a commercially available digital circuit for each circuit in figure 109.

2-input NAND 00, 2-input nor 02, 3-input nor 27, not 04, $D_{FF}$ 74, 3 to 8 decoder 138,

**7** What is the set of datapath inputs that replace M by N?

sNx, sxz, ldM

**8** What is the set of datapath inputs that replace N by R?

sRx, sxz, ldN

**9** What is the set of datapath inputs that divide M by N and store the result in R?

sNx, sMy, sdz, ldR

**Chapter 1**

| $x$ | $y$ | $f101$ $xy'$ | $f102$ $x+y'$ | $f103$ $x \oplus y'$ | $f104$ $(x+y)(x+y')$ | $f105$ $x+xy$ | $f106$ $x(x+y)$ |
|---|---|---|---|---|---|---|---|
| 0 | 0 | 0 | 1 | 1 | 0 | 0 | 0 |
| 0 | 1 | 0 | 0 | 0 | 0 | 0 | 0 |
| 1 | 0 | 1 | 1 | 0 | 1 | 1 | 1 |
| 1 | 1 | 0 | 1 | 1 | 1 | 1 | 1 |

| $x$ | $y$ | $z$ | $f107$ $(x+y)(x'+z')(y+z)$ | $f108$ $xy+x'z+yz$ |
|---|---|---|---|---|
| 0 | 0 | 0 | 0 | 0 |
| 0 | 0 | 1 | 0 | 1 |
| 0 | 1 | 0 | 1 | 0 |
| 0 | 1 | 1 | 1 | 1 |
| 1 | 0 | 0 | 0 | 0 |
| 1 | 0 | 1 | 0 | 0 |
| 1 | 1 | 0 | 0 | 1 |
| 1 | 1 | 1 | 0 | 1 |

117 *mux* $f(z) = g(za_1 + z'a_0)$

118 *mux* $f(y,z) = g(yza_3 + yz'a_2 + y'za_1 + y'z'a_0)$

119 *mux* $f(x,y,z) = g(xyza_7 + xyz'a_6 + xy'za_5 + xy'z'a_4$
$+ x'yza_3 + x'yz'a_2 + x'y'za_1 + x'y'z'a_0)$

| $x$ | $y$ | $z$ | $f117$ $(za_1 + z'a_0)$ | $f118$ $(yza_3 + yz'a_2 + y'za_1 + y'z'a_0)$ | $f119$ $eqn119$ |
|---|---|---|---|---|---|
| 0 | 0 | 0 | $a_0$ | $a_0$ | $a_0$ |
| 0 | 0 | 1 | $a_1$ | $a_1$ | $a_1$ |
| 0 | 1 | 0 | | $a_2$ | $a_2$ |
| 0 | 1 | 1 | | $a_3$ | $a_3$ |
| 1 | 0 | 0 | | | $a_4$ |
| 1 | 0 | 1 | | | $a_5$ |
| 1 | 1 | 0 | | | $a_6$ |
| 1 | 1 | 1 | | | $a_7$ |

---

**201-204 Since g=f′ the g are zeros and the f are ones on the maps.**

$201\ f = (yz)'$ $\qquad\qquad$ $g = yz$

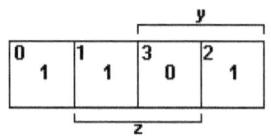

$202\ f = y + z$ $\qquad\qquad$ $g = y'z'$

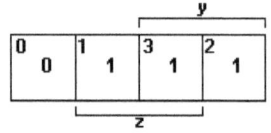

$203\ f = y\ xor\ z$ $\qquad\qquad$ $\underline{g = y\ xnor\ z}$

$204\ f = xy + x'z + yz$ $\qquad$ $g = (x' + y')(x + z')(y' + z')$

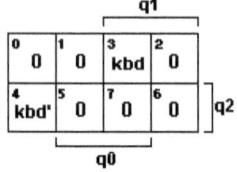 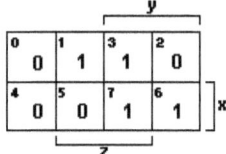

$205\ \ d_2 = kbd \times S_3 + kbd' \times S_4$ (from page 10)

$206\ \ d_1 = S_1 + S_3 \times kbd'$

$207\ \ d_0 = S_0 \times (R < N) \times (R = 0)' + S_3 \cdot kbd'$

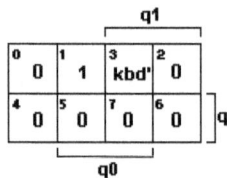 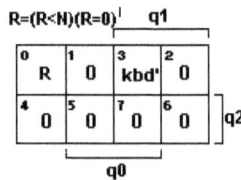

$208\ d_0 = q_0\ xor\ p$ $\qquad$ $209\ d_1 = q_1\ xor\ pq_0$ $\qquad$ $210\ d_2 = q_2\ xor\ pq_1q_0$

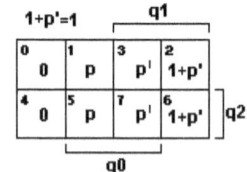 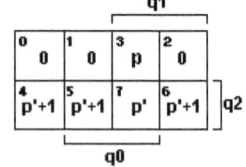

$211\ f = z(z' + y) = zz' + yz = yz$

**212** $f = xy + x'z + yz$

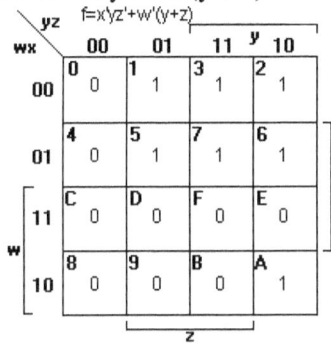

**213** $f = (x + y)(x' + z)(y + z)$

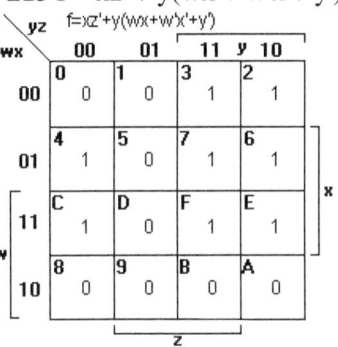

**214** $f = x'yz' + w'(y + z)$

**215** $f = xz' + y(wx + w'x' + y')$

**216** *mux* $f(y, z) = yz a_3 + yz' a_2 + y' z a_1 + y' z' a_0$     (mev $a_0$ to $a_3$)

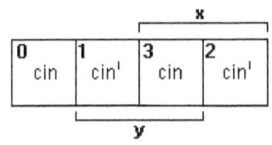

**217** *Full adder* $s = x \oplus y \oplus c_{in}$   $c = xy + c_{in}(x + y)$   (mev $c_{in}$)

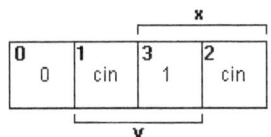

**218** $d_2 = q_2 \oplus p q_1 q_0$   (mev p)

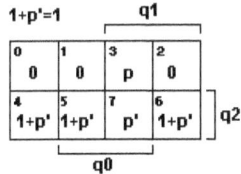

235

**413**

$f = (x \oplus y)z = x'yz + xy'z = x'm_3 + xm_1$

*address inputs* $y = b,\ z = a$

*data inputs* $d_0 = L \quad d_1 = x \quad d_2 = L \quad d_3 = x'$

**414**

$f = xyz + (x \oplus y)z = xyz + x'yz + xy'z = xm_3 + x'm_3 + xm_1 = m_3 + xm_1$

*address inputs* $y = b,\ z = a$

*data inputs* $d_0 = L \quad d_1 = x \quad d_2 = L \quad d_3 = H$

**419**

$f = xy + (x \oplus y)z = xy(z + z') + x'yz + xy'z = m_7 + m_6 + m_3 + m_5$

*address inputs* $x = c,\ y = b,\ z = a$

*data inputs* $d_3 = d_5 = d_6 = d_7 = H,\quad d_0 = d_1 = d_2 = d_4 = L$

**421**

$f = wx + y'z = wx(y + y')(z + z') + (x + x')y'z$

$= w[m_7 + m_6 + m_5 + m_4] + m_5 + m_1 = w[m_7 + m_6 + m_4] + m_5 + m_1$

*address inputs* $x = c,\ y = b,\ z = a$

*data inputs* $d_4 = d_6 = d_7 = w,\quad d_1 = d_5 = H,\quad d_0 = d_2 = d_3 = L$

423   $f = (x \oplus y)z = x'yz + xy'z = m_3 + m_5$

*data inputs* $d_3 = d_5 = H,\quad all\ others = L$

424   $f = xyz + (x \oplus y)z = xyz + x'yz + xy'z = m_7 + m_3 + m_5$

*data inputs* $d_3 = d_5 = d_7 = H,\quad all\ others = L$

425   $f = xy + xz + yz = xy(z + z') + x(y + y')z + (x + x')yz = m_7 + m_6 + m_5 + m_3$

*data inputs* $d_3 = d_5 = d_6 = d_7 = H,\quad all\ others = L$

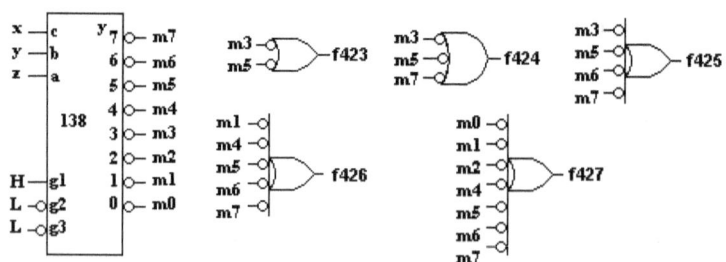

**428** Make a table of addresses. Address lines that do not change are inputs to an AND connected to the decoder enable input. Lines $a_5a_4a_3$ range from 0 to 7 so they are connected to the decoder's cba address inputs.

| addr | $a_7a_6$ | $a_5a_4a_3$ | $a_2a_1a_0$ |
|------|------|------|------|
| 87 | 10 | 000 | 111 |
| 8F | 10 | 001 | 111 |
| 97 | 10 | 010 | 111 |
| 9F | 10 | 011 | 111 |
| A7 | 10 | 100 | 111 |
| AF | 10 | 101 | 111 |
| B7 | 10 | 110 | 111 |
| BF | 10 | 111 | 111 |

**430-431** Make a table of codes. Include a minterm in a y expression when a y=1.

$y_3=w$

$y_2=m4+m5+m6+m7+m8+m9+mA+mB$

$=w'$ [m4+m5+m6+m7] +w [m0+m1+m2+m3]

$=w'$ [x (m0+m1+m2+m3)] +w [x' (m0+m1+m2+m3)]

$y_1=m2+m3+m4+m5+m8+m9+mE+mF$

$= w'$ [m2+m3+m4+m5]+w [m0+m1+m6+m7]

$= w'$ [x' (m2+m3)+x (m0+m1)]

$\qquad +w$ [x' (m0+m1)+x' (m2+m3]

$y_0=m1+m2+m4+m7+m8+mB+mD+mE$

$= w'$ [m1+m2+m4+m7]+w [m0+m3+m5+m6]

$= w'$ [x' (m1+m2)+ x (m0+m3)]

$\qquad +w$ [x' (m0+m3)+ x (m1+m2)]

| wxyz | $y_3y_2y_1y_0$ |
|------|------|
| 0000 | 0000 |
| 0001 | 0001 |
| 0011 | 0010 |
| 0010 | 0011 |
| 0110 | 0100 |
| 0111 | 0101 |
| 0101 | 0110 |
| 0100 | 0111 |
| 1100 | 1000 |
| 1101 | 1001 |
| 1111 | 1010 |
| 1110 | 1011 |
| 1010 | 1100 |
| 1011 | 1101 |
| 1001 | 1110 |
| 1000 | 1111 |

**430**

xyz are cba address inputs to three 8:1 mux for $y_0$, $y_1$, $y_2$

$y_3 = w$

$y_2$ mux -- w' is connected to inputs 4,5,6,7 and w to inputs 0,1,2,3.

$y_1$ mux -- w' is connected to inputs 2,3,4,5 and w to inputs 0,1,6,7.

$y_0$ mux -- w' is connected to inputs 1,2,4,7 and w to inputs 0,3,5,6.

## 431

yz are ba address inputs to six 4:1 mux for $y_0$, $y_1$, $y_2$.
$y_3 = w$

$y_2 = y_{21}$ mux OR $y_{22}$ mux.
$y_{21}$ mux – $w'$ enables, x to inputs 0,1,2,3
$y_{22}$ mux – w enables, $x'$ to inputs 0,1,2,3

$y_1 = y_{11}$ mux OR $y_{12}$ mux.
$y_{11}$ mux – $w'$ enables, x to inputs 0,1 and $x'$ to inputs 2,3.
$y_{12}$ mux – w enables, $x'$ to inputs 0,1 and x to inputs 2,3

$y_0 = y_{01}$ mux OR $y_{02}$ mux.
$y_{01}$ mux – $w'$ enables, x to inputs 0,3 and $x'$ to inputs 1,2.
$y_{12}$ mux – w enables, $x'$ to inputs 0,3 and x to inputs 1,2.

## 432

$$c_4 = g_4 + p_4 c_3$$
$$= g_4 + p_4(g_3 + p_3 g_2 + p_3 p_2 g_1 + p_3 p_2 p_1 g_0 + p_3 p_2 p_1 p_0 c_{in})$$

## 433

$$c_1 = g_1 + p_1 g_0 + p_1 p_0 c_{in}$$
$$= x_1 y_1 + (x_1 + y_1)x_0 y_0 + (x_1 + y_1)(x_0 + y_0)c_{in}$$

## 434

$$g_k p_k = (s_3 ba + s_2 b'a)(s_1 b' + s_0 b + a)$$
$$= s_3 ba(s_1 b' + s_0 b + a) + s_2 b'a(s_1 b' + s_0 b + a)$$
$$= (s_3 bas_1 b' + s_3 bas_0 b + s_3 baa) + (s_2 b'as_1 b' + s_2 b'as_0 b + s_2 b'aa)$$
$$= (0 + s_3 ba + s_3 bas_0 + s_3 ba) + (0 + 0 + s_2 b'a) = s_3 ba(s_0 + 1) + s_2 b'a$$
$$= s_3 ba + s_2 b'a$$
$$= g_k$$

## 435   s=0100

$$g_4 p_4 = (0 \cdot ba + 1 \cdot b'a)(0 \cdot b' + 0 \cdot b + a) = (b'a)(a) = b'a = g_4$$

**436**   s=0110

$$z_k = g_k \oplus p_k = (s_3 ba + s_2 b'a) \oplus (s_1 b' + s_0 b + a)$$
$$z_6 = g_6 \oplus p_6 = (0 \cdot ba + 1 \cdot b'a) \oplus (1 \cdot b' + 0 \cdot b + a)$$
$$= b'a \oplus (b' + a) = b'a(b' + a)' + (b'a)'(b' + a)$$
$$= b'aba' + (b + a')(b' + a) = bb' + ba + a'b' + a'a$$
$$= (b')'a + a'(b') = b' \oplus a$$

$\quad$ *or*

$$= b(a')' + (a')b' = b \oplus a'$$

**437**

$$s = x \oplus y = xx' + xy' + x'y + yy' = (x' + y')(x + y) = (xy)'(x + y) = g'p$$

**438**   s=1001

$$z_9 = g_9 \oplus p_9 = (1 \cdot ba + 0 \cdot b'a) \oplus (0 \cdot b' + 1 \cdot b + a)$$
$$= ba \oplus (b + a) = ba(b + a)' + (ba)'(b + a) = ba(b'a') + (b' + a')(b + a)$$
$$= b'b + b'a + a'b + a'a = b'a + a'b = b \oplus a$$

**440**

For k use code E to get k=$z_E$=b'a'.
For p use code 1 to get p=$z_1$=b+a.
For s use code 9 to get s=$z_9$=b xor a.

**441** If k is H (true) then mn is turned on grounding the carry line so that c=0. If g is L (true) the mp pulls the carry line to H so that c=1. If p is thru the TG is turned on so that $c_n = p_n c_{n-1}$. Therefore $c_n = k_n'(g_n + p_n c_{n-1})$.

**442**

$\quad$ *if* $\ x = x_1 x_0\ $ *and* $y = y_1 y_0,$
$\quad$ *then* $x = y$ *when* $x_1 = y_1\ $ *and* $x_0 = y_0$
$\quad$ *now* $x_k = y_k$ *when* $(x_k \oplus y_k)' = 1$
$\quad$ *then* $x = y$ *when* $\quad f = (x_1 \oplus y_1)'(x_0 \oplus y_0)' = 1$

**443 Unsigned x and y**

$\quad$ *if* $\ x = x_1 x_0\ $ *and* $y = y_1 y_0,$
$\quad$ *then* $x > y$ *when* $(x_1 > y_1)$ *or* $\ (x_1 = y_1$ *and* $x_0 > y_0)$
$\quad$ *now* $x_k > y_k$ *when* $x_k y_k' = 1$
$\quad$ *then* $x > y$ *when* $\quad f = (x_1 y_1') + (x_1 \oplus y_1)'(x_0 y_0') = 1$

**439** Make a table of codes where wxyz is binary input, and $f_4$ is the tens BCD digit and $f_{3210}$ are the units BCD digit. Then make an $f_4$ K map. From the map $f_4=wx+wy$. Two ANDs and 1 one OR generate $f_4$.

When $f_4=1$ add 6 to wxyz to create 10000 to 10110. Discard high bit to get $f_{3210} = 0000$ to 0110.

Implement code converter by connecting $f_4$ to mux control input to select 0000 or 0110 as inputs to alu A inputs. Connect wxyz to alu B inputs.

Connect ALU $c_{in}$ to L

| wxyz | $f_4$ | $f_3f_2f_1f_0$ |
|------|-------|-------|
| 0000 | 0 | 0000 |
| 0001 | 0 | 0001 |
| 0010 | 0 | 0010 |
| 0011 | 0 | 0011 |
| 0100 | 0 | 0100 |
| 0101 | 0 | 0101 |
| 0110 | 0 | 0110 |
| 0111 | 0 | 0111 |
| 1000 | 0 | 1000 |
| 1001 | 0 | 1001 |
| 1010 | 1 | 0000 |
| 1011 | 1 | 0001 |
| 1100 | 1 | 0010 |
| 1101 | 1 | 0011 |
| 1110 | 1 | 0100 |
| 1111 | 1 | 0101 |

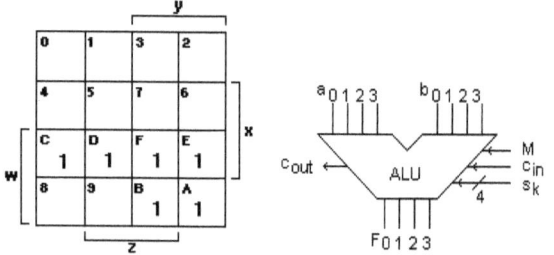

**503**

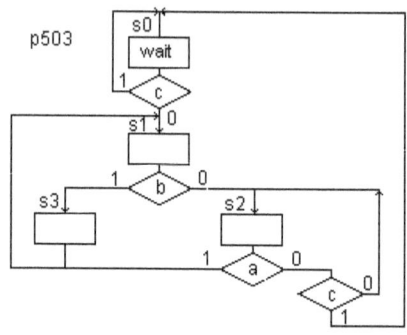

| Truth Table for JK and D Flip-flop Excitation Tables | | | |
|------|------|------|------|
| Transition | j | k | d |
| 0 to 0 | 0 | - | 0 |
| 0 to 1 | 1 | - | 1 |
| 1 to 0 | - | 1 | 0 |
| 1 to 1 | - | 0 | 1 |

| PS $q_1q_0$ | a | b | c | NS $q_1^+q_0^+$ | wait | $j_1$ | $k_1$ | $j_0$ | $k_0$ | $d_1$ | $d_0$ |
|---|---|---|---|---|---|---|---|---|---|---|---|
| 00 | − | − | 0 | 01 | 1 | 0 | − | 1 | − | 0 | 1 |
| 00 | − | − | 1 | 00 | 1 | 0 | − | 0 | − | 0 | 0 |
| 01 | − | 0 | − | 10 | 0 | 1 | − | − | 1 | 1 | 0 |
| 01 | − | 1 | − | 11 | 0 | 1 | − | − | 0 | 1 | 1 |
| 10 | 0 | − | 0 | 10 | 0 | − | 0 | 0 | − | 1 | 0 |
| 10 | 0 | − | 1 | 00 | 0 | − | 1 | 0 | − | 0 | 0 |
| 10 | 1 | − | − | 01 | 0 | − | 1 | 1 | − | 0 | 1 |
| 11 | − | − | − | 01 | 0 | − | 1 | − | 0 | 0 | 1 |

**501**

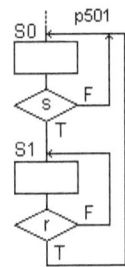

$S_0 = q_0'$   $S_1 = q_0$

d ff   $d_0 = sq_0' + r'q_0$

jk ff   $j_0 = sq_0'$   $k_0 = rq_0$

| PS $q_0$ | s | r | NS $q_0^+$ | $d_0$ | $j_0$ | $k_0$ |
|---|---|---|---|---|---|---|
| 0 | 0 | − | 0 | 0 | 0 | − |
| 0 | 1 | − | 1 | 1 | 1 | − |
| 1 | − | 0 | 1 | 1 | − | 0 |
| 1 | − | 1 | 0 | 0 | − | 1 |

**502**

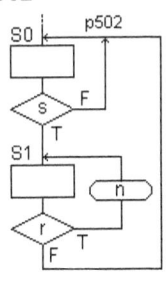

d ff $d_0 = sq_0' + rq_0$

jk ff $j = sq_0'$

$k = r'q_0$

$n = rq_0$

| PS $q_0$ | s | r | NS $q_0^+$ | n | $d_0$ | $j_0$ | $k_0$ |
|---|---|---|---|---|---|---|---|
| 0 | 0 | − | 0 | 0 | 0 | 0 | − |
| 0 | 1 | − | 1 | 0 | 1 | 1 | − |
| 1 | − | 0 | 0 | 0 | 0 | − | 1 |
| 1 | − | 1 | 1 | 1 | 1 | − | 0 |

**504**

| Truth Table p504, PS, PI, NS, d inputs to D Flip-flops | | | | | | | | |
|---|---|---|---|---|---|---|---|---|
| PS= $q_2q_1q_0$ | PI C | PI b | PI 16 | PI f | NS= $q_2^+q_1^+q_0^+$ | d2 | d1 | d0 |
| 000 | 0 | - | - | - | 001 | 0 | 0 | 1 |
| 000 | 1 | - | - | - | 000 | 0 | 0 | 0 |
| 001 | - | 0 | - | - | 011 | 0 | 1 | 1 |
| 001 | - | 1 | - | - | 101 | 1 | 0 | 1 |
| 010 | - | - | - | - | 001 | 0 | 0 | 1 |
| 011 | 0 | - | 0 | - | 011 | 0 | 1 | 1 |
| 011 | 1 | - | 0 | - | 000 | 0 | 0 | 0 |
| 011 | - | - | 1 | 0 | 111 | 1 | 1 | 1 |
| 011 | - | - | 1 | 1 | 010 | 0 | 1 | 0 |
| 101 | - | - | - | - | 001 | 0 | 0 | 1 |
| 111 | 0 | - | - | - | 111 | 1 | 1 | 1 |
| 111 | 1 | - | - | - | 000 | 0 | 0 | 0 |

**505** sequence of states 0132013 etc

| PS | NS |
|---|---|
| $q_1q_0$ | $q_1^+q_0^+$ |
| 00 | 01 |
| 01 | 11 |
| 10 | 00 |
| 11 | 10 |

504

| c | 11000000010 |
|---|---|
| b | 00000100000 |
| a | 11111100001 |
| State | 21212131220 |

**506 If** in=1=H, then possible state sequences are 01374250 etc., and (stuck on) 6.

| $q_2q_1q_0$ | $d_2=q_1$ | $d_1=q_2 \oplus q_0$ | $d_0=q_2 \oplus in$ | $q_2^+q_1^+q_0^+$ |
|---|---|---|---|---|
| 000 | 0 | 0 | 1 | 001 |
| 001 | 0 | 1 | 1 | 011 |
| 010 | 1 | 0 | 1 | 101 |
| 011 | 1 | 1 | 1 | 111 |
| 100 | 0 | 1 | 0 | 010 |
| 101 | 0 | 0 | 0 | 000 |
| 110 | 1 | 1 | 0 | 110 |
| 111 | 1 | 0 | 0 | 100 |

**507**

Mainline

| $Q_1q_0$ | s | n | $q_1^+q_0^+$ | Sub |
|---|---|---|---|---|
| 00 | 0 | - | 00 | 0 |
| 00 | 1 | - | 01 | 0 |
| 01 | - | 0 | 01 | 1 |
| 01 | - | 1 | 11 | 1 |
| 10 | - | - | 00 | 0 |
| 11 | - | - | 10 | 0 |

Link

| $q_1q_0$ | sub | r | $q_1^+q_0^+$ | n |
|---|---|---|---|---|
| 00 | 0 | - | 00 | 0 |
| 00 | 1 | - | 01 | 0 |
| 01 | - | 0 | 01 | 0 |
| 01 | - | 1 | 11 | 0 |
| 10 | - | - | 00 | 1 |
| 11 | - | - | 10 | 0 |

**508**

| $q_0$ | R | $q_0^+$ | count | red | load7 |
|---|---|---|---|---|---|
| 0 | 0 | 0 | 1 | 0 | 0 |
| 0 | 1 | 1 | 1 | 0 | 0 |
| 1 | - | 0 | 0 | 1 | 1 |

**509**

| $q_0$ | R | $q_0^+$ | count | red | load7 |
|---|---|---|---|---|---|
| 0 | 0 | 0 | 1 | 0 | 0 |
| 0 | 1 | 1 | 1 | 0 | 1 |
| 1 | - | 0 | 0 | 1 | 0 |

**601**

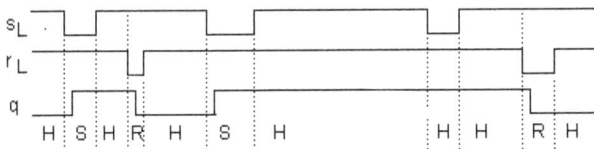

**602**

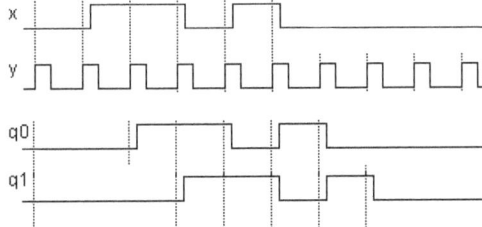

**604**

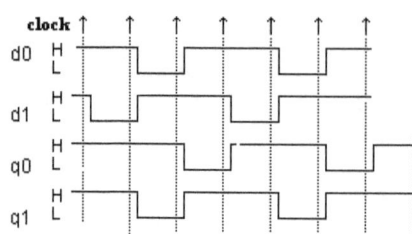

$d_0 = q_1$, $d_1 = q_1 \oplus q_0$, $H = 1$

| clock | $q_1$ | $q_0$ | $d_1$ | $d_0$ | $q_1^+$ | $q_0^+$ |
|---|---|---|---|---|---|---|
| $n+0$ | H | H | L | H | L | H |
| $n+1$ | L | H | H | L | H | L |
| $n+2$ | H | L | H | H | H | H |
| $n+3$ | H | H | L | H | L | H |
| $n+4$ | L | H | H | L | H | L |
| $n+5$ | H | L | H | H | H | H |

## 605

$d_0 = q_2'$, $d_1 = q_0$, $d_2 = q_2' \oplus q_1$, $H = 1$

| clock | $q_2$ | $q_1$ | $q_0$ | $d_2$ | $d_1$ | $d_0$ | $q_2^+$ | $q_1^+$ | $q_0^+$ |
|---|---|---|---|---|---|---|---|---|---|
| $n+0$ | H | H | H | H | H | L | H | H | L |
| $n+1$ | H | H | L | H | L | L | H | L | L |
| $n+2$ | H | L | L | L | L | L | L | L | L |
| $n+3$ | L | L | L | H | L | H | H | L | H |
| $n+4$ | H | L | H | L | H | L | L | H | L |
| $n+5$ | L | H | L | L | L | H | L | L | H |
| $n+6$ | L | L | H | H | H | H | H | H | H |
| $n+7$ | H | H | H | H | H | L | H | H | L |
| $n+8$ | H | H | L | H | L | L | H | L | L |

## 606

$q_0^+ = d_0 = x$

$q_1^+ = d_1 = q_0$

| PS | PI | NS |
|---|---|---|
| $q_1q_0$ | $x$ | $q_1^+q_0^+$ |
| 00 | 0 | 00 |
| 00 | 1 | 01 |
| 01 | 0 | 10 |
| 01 | 1 | 11 |
| 10 | 0 | 00 |
| 10 | 1 | 01 |
| 11 | 0 | 10 |
| 11 | 1 | 11 |

## 607

$q_0^+ = d_0 = q_1'$

$q_1^+ = d_1 = q_0$

| PS | NS |
|---|---|
| $q_1q_0$ | $q_1^+q_0^+$ |
| 00 | 01 |
| 01 | 11 |
| 10 | 00 |
| 11 | 10 |

## 608

$q_0^+ = d_0 = q_1$

$q_1^+ = d_1 = q_1 \oplus q_0$

| PS | NS |
|---|---|
| $q_1q_0$ | $q_1^+q_0^+$ |
| 00 | 00 |
| 01 | 10 |
| 10 | 11 |
| 11 | 01 |

**609**

| PS | NS |
|---|---|
| $q_2 q_1 q_0$ | $q_2^+ q_1^+ q_0^+$ |
| 000 | 101 |
| 001 | 111 |
| 010 | 001 |
| 011 | 011 |
| 100 | 000 |
| 101 | 010 |
| 110 | 100 |
| 111 | 110 |

$q_0^+ = d_0 = q_2'$

$q_1^+ = d_1 = q_0$

$q_2^+ = d_2 = q_2' \oplus q_1$

state seq is 05217640 (no 3)

**611**

$j_0 = k_0 = 1, \ j_1 = k_1 = q_0'$

$q_0^+ = j_0 q_0' + k_0' q_0 = 1 q_0' + 0 q_0 = q_0'$

$q_1^+ = j_1 q_1' + k_1' q_1 = q_0' q_1' + q_0 q_1$

| PS | NS |
|---|---|
| $q_1 q_0$ | $q_1^+ q_0^+$ |
| 00 | 11 |
| 01 | 00 |
| 10 | 01 |
| 11 | 10 |

**612**

$q_0^+ = j_0 q_0' + k_0' q_0$

$= (q_1 \oplus q_0') q_0' + (q_1 \oplus q_0')' q_0$

$= [q_1'(q_0') + q_1(q_0')'] q_0' + [q_1'(q_0') + q_1(q_0')'] q_0$

$= [q_1' q_0' + 0] + [0 + q_1 q_0] = q_1' q_0' + q_1 q_0$

$q_1^+ = j_1 q_1' + k_1' q_1 = q_0 q_1' + q_0 q_1 = q_0$

| PS | NS |
|---|---|
| $q_1 q_0$ | $q_1^+ q_0^+$ |
| 00 | 01 |
| 01 | 10 |
| 10 | 00 |
| 11 | 11 |

**613**

$j_0 = k_0 = 1, \ j_1 = k_1 = q_0, \ j_2 = k_2 = q_1 q_0$

$q_0^+ = 1 q_0' + 1' q_0 = q_0' = q_0 \oplus 1$

$q_1^+ = q_0 q_1' + q_0' q_1 = q_1 \oplus q_0$

$q_2^+ = (q_1 q_0) q_2' + (q_1 q_0)' q_2 = q_2 \oplus q_1 q_0$

| PS | NS |
|---|---|
| $q_2 q_1 q_0$ | $q_2^+ q_1^+ q_0^+$ |
| 000 | 001 |
| 001 | 010 |
| 010 | 011 |
| 011 | 100 |
| 100 | 101 |
| 101 | 110 |
| 110 | 111 |
| 111 | 000 |

**614 see 611      615 see 612      616 see 613**
**617** Connect t input to j and k FF inputs.
**618**   $q^+ = q$ xor t   therefore

$q_0{}^+ = q_0$ xor 1          $\Rightarrow$      $t = 1$
$q_1{}^+ = q_1$ xor $q_0{}'$        $\Rightarrow$      $t = q_0{}'$
$q_2{}^+ = q_2$ xor $q_1{}'q_0{}'$      $\Rightarrow$      $t = q_1{}'q_0{}'$
$q_3{}^+ = q_3$ xor $q_2{}'q_1{}'q_0{}'$   $\Rightarrow$      $t = q_2{}'q_1{}'q_0{}'$

**619**
$q^+ = q' = q$ xor 1     $\Rightarrow$      $t = 1$

**620**
$q_0{}^+ = q_0{}' = q_0 \oplus 1$  $\Rightarrow t = H$

$q_1{}^+ = q_1 \oplus q_0$     $\Rightarrow t = q_0$

---

**701 Problem 701** Use the up counter $d_2$, $d_1$, $d_0$ equations above to draw K maps with $q_2{}^+$, $q_1{}^+$, $q_0{}^+$ as coordinates. Then from the K maps derive new $d_2$, $d_1$, $d_0$ equations in xor format. Compare to the algebra on page 130.

$d_0 = s_0 + s_2 + s_4 + s_6$
$d_1 = s_1 + s_2 + s_5 + s_6$
$d_2 = s_3 + s_4 + s_5 + s_6$

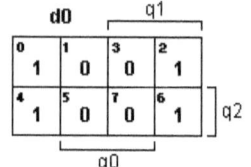 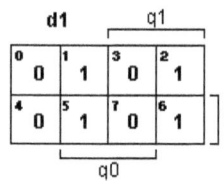 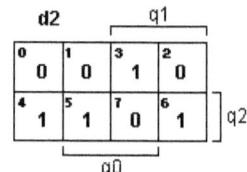

A direct take from the maps is as follows
$d_0 = (m_0 + m_4) + (m_2 + m_6) = q_1{}'q_0{}' + q_1 q_0{}' = q_0{}' = q_0 \oplus 1$   $\rightarrow$   $d_0 = q_0 \oplus p$
$d_1 = (m_1 + m_5) + (m_2 + m_6) = q_1{}'q_0 + q_1 q_0{}' = q_1 \oplus q_0$   $\rightarrow$   $d_1 = q_1 \oplus pq_0$
$d_0 = (m_3) + (m_4 + m_5 + m_6) = q_2{}'q_1 q_0 + q_2(q_1{}'q_0{}' + q_1{}'q_0 + q_1 q_0{}')$
$\quad = q_2{}'q_1 q_0 + q_2(q_1{}'[q_0{}' + q_0] + q_1 q_0{}') = q_2{}'q_1 q_0 + q_2(q_1 q_0)'$
$\quad = q_2 \oplus q_1 q_0 \rightarrow d_2 = q_2 \oplus pq_1 q_0$

**703** Count control p holds counter in present state or advances counter to next state. Count seq is 3210321....

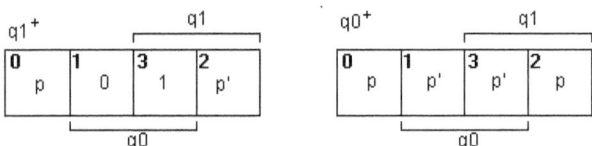

$$q_1^+ = pm_0 + 0m_1 + p'm_2 + m_3$$
$$= pq_1'q_0' + q_1q_0 + p'q_1q_0'$$
$$= q_1'(pq_0') + q_1(p'q_0' + q_0)$$
$$= q_1'(pq_0') + q_1(p' + q_0)$$
$$= q_1'(pq_0') + q_1(pq_0')' = q_1 \oplus (pq_0')$$
$$q_0^+ = pm_0 + p'm_1 + pm_2 + p'm_3$$
$$= pq_1'q_0' + p'q_1'q_0 + p'q_1q_0' + pq_1q_0'$$
$$= pq_0'(q_1' + q_1) + p'q_0(q_1' + q_1)$$
$$= pq_0' + p'q_0 = q_0 \oplus (p)$$

| $q_1q_0$ | p | $q_1^+q_0^+$ |
|---|---|---|
| 00 | 0 | 00 |
| 00 | 1 | 11 |
| 01 | 0 | 01 |
| 01 | 1 | 00 |
| 10 | 0 | 10 |
| 10 | 1 | 01 |
| 11 | 0 | 11 |
| 11 | 1 | 10 |

**704** Count control p holds counter in present state or advances counter to next state. Count seq is 0123012.... or 3210321....

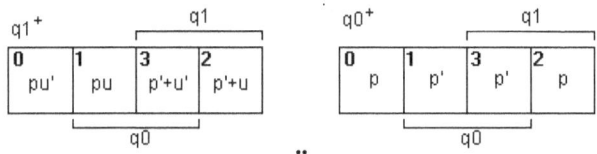

$$q_1^+ = pu'm_0 + pum_1 + (p' + pu)m_2 + (p' + pu')m_3$$
$$= pu'q_1'q_0' + puq_1'q_0 + (p'+u')q_1q_0 + (p'+u)q_1q_0'$$
$$= q_1'(pu'q_0' + puq_0) + q_1[(p'+u')q_0 + (p'+u)q_0']$$
$$= q_1'p(u'q_0' + uq_0) + q_1[p'(q_0 + q_0') + (u'q_0 + uq_0')]$$
$$= q_1'p(u'q_0' + uq_0) + q_1[p' + (u'q_0 + uq_0')]$$
$$= q_1'p(u'q_0' + uq_0) + q_1[p(uq_0 + u'q_0')]'$$
$$= q_1 \oplus p(u'q_0' + uq_0)$$
$$q_0^+ = pm_0 + p'm_1 + pm_2 + p'm_3$$
$$= q_0 \oplus (p)$$

| $q_1q_0$ | p | u | $q_1^+q_0^+$ |
|---|---|---|---|
| 00 | 0 | - | 00 |
| 00 | 1 | 0 | 11 |
| 00 | 1 | 1 | 01 |
| 01 | 0 | - | 01 |
| 01 | 1 | 0 | 00 |
| 01 | 1 | 1 | 10 |
| 10 | 0 | - | 10 |
| 10 | 1 | 0 | 01 |
| 10 | 1 | 1 | 11 |
| 11 | 0 | - | 11 |
| 11 | 1 | 0 | 10 |
| 11 | 1 | 1 | 00 |

Digital Design

**706**

| $q_2 q_1 q_0$ | p | $q_2^+ q_1^+ q_0^+$ |
|---|---|---|
| 000 | 0 | 000 |
| 000 | 1 | 001 |
| 001 | 0 | 001 |
| 001 | 1 | 010 |
| 010 | 0 | 010 |
| 010 | 1 | 011 |
| 011 | 0 | 011 |
| 011 | 1 | 100 |
| 100 | 0 | 100 |
| 100 | 1 | 101 |
| 101 | 0 | 101 |
| 101 | 1 | 110 |
| 110 | 0 | 110 |
| 110 | 1 | 111 |
| 111 | 0 | 111 |
| 111 | 1 | 000 |

Make the maps from the table entries.

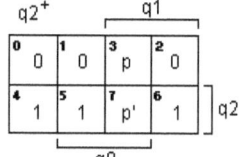 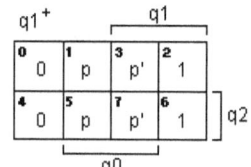 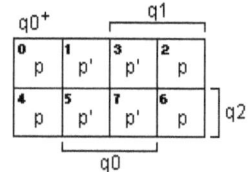

We make several observations to acquire insight, and save time and energy. Clearly $q_0$ does not depend on $q_1$ and $q_2$, and $q_1$ does not depend on $q_2$. The $q_0^+$ map has two rows that are the same as the 701 $q_0^+$ map. Thus each row's equation is $q_0$ xor p. In this $q_0^+$ map rows are multiplied by $q_2'$ and $q_2$. The same comments apply to the $q_1$ map. And so

$$q_0^+ = q_2'(q_0 \oplus p) + q_2(q_0 \oplus p) = q_0 \oplus p$$

$$q_1^+ = q_2'(q_1 \oplus pq_0) + q_2(q_1 \oplus pq_0) = q_1 \oplus pq_0$$

$$q_2^+ = q_2'(pq_1q_0) + q_2(q_1'q_0' + q_1'q_0 + p'q_1q_0 + q_1q_0')$$

$$= q_2'(pq_1q_0) + q_2(q_0'[q_1' + q_1] + q_0[q_1' + p'q_1])$$

$$= q_2'(pq_1q_0) + q_2(q_0' + q_0[q_1' + p'])$$

$$= q_2'(pq_1q_0) + q_2(q_0' + q_1' + p')$$

$$= q_2'(pq_1q_0) + q_2(pq_1q_0)' = q_2 \oplus pq_1q_0$$

248

**707** Produce 1111000000000. The 13 clock periods require the counter to count up to $1010_2$ that produces decoded output $n_A$. The 4 ones require counting C, D, E, F in order to use the 163's state $F_{16}$ decoder that produces decoded output $n_F$. Thus active low $n_A$ loads $1100_2$ into the counter, and is connected to the k input of a jk flip-flop to change the active low q to H (w=H). Active high $n_F$ is connected to the j input to change the active low q to L (w=L). State 0 follows F so there is no need to clear the counter to 0.

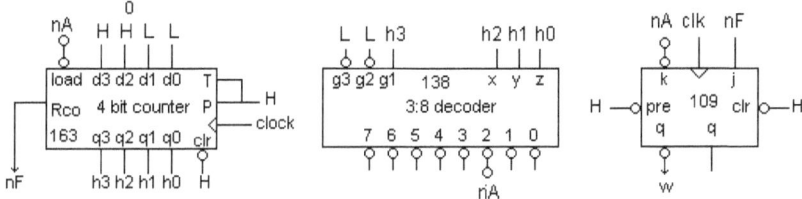

**708** State sequence is FEDCBA98.

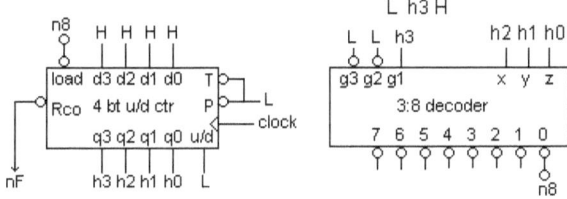

**709** Design a modulo $37_{10}$ up counter with states 0 to $36_{10}$ sequence.

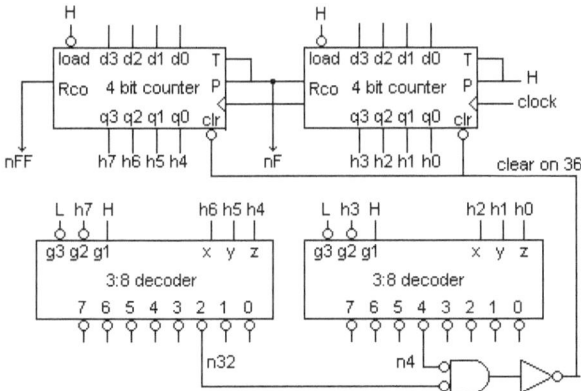

**710** The count from 00 to FF requires two 163 chips. Active high $n_{FF}$ is the second 163 $R_{CO}$ output. Use $n_{FF}$ to produce active low $n_K$ when w is H ($n_K = n_{FF}$ w). The $n_K$ waveform clears the counter and sets w to L. Since 00 follows FF clearing is not necessary. Also use $n_{FF}$ to produce active high $n_J$ when w is L ($n_J = n_{FF}$ w'). The $n_J$ waveform loads F1 into the counter and sets w to H.

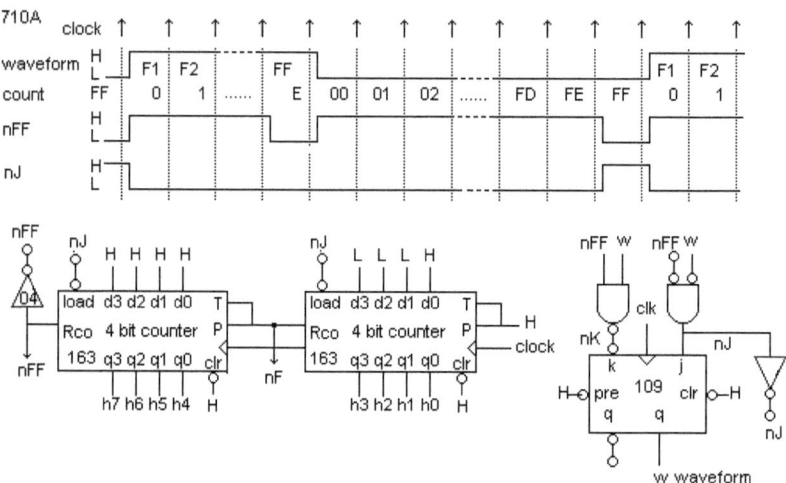

**711** The count from 03 to 9E requires two 163 chips. Use $n_{0C}$ to produce active low $n_K$ when w is 1 ($n_K = n_{0C}$ w). The $n_K$ waveform loads 03 into the counter and sets w to L. The $n_{9E}$ produces active high $n_J$ when w is L ($n_J = n_{9E}$ w'). The $n_J$ waveform clears the counter and sets w to H.

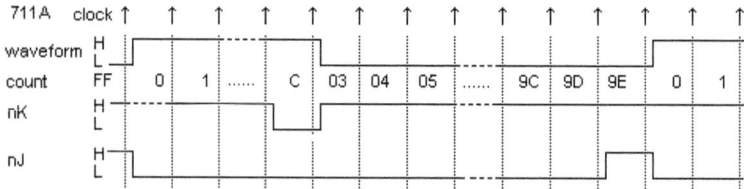

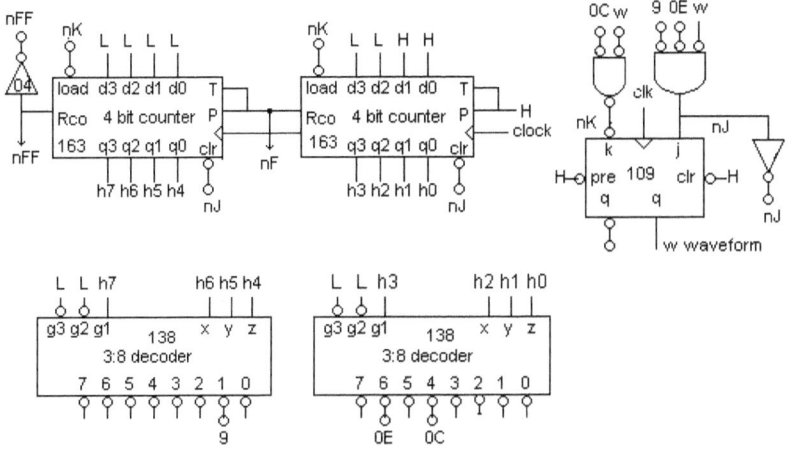

## 712

| PS | $q_2q_1q_0$ | $q_2^+q_1^+q_0^+$ | NS |
|---|---|---|---|
| $s_0$ | 000 | 000 | $s_0$ |
| $s_1$ | 001 | 000 | $s_0$ |
| $s_2$ | 010 | 001 | $s_1$ |
| $s_3$ | 011 | 001 | $s_1$ |
| $s_4$ | 100 | 010 | $s_2$ |
| $s_5$ | 101 | 010 | $s_2$ |
| $s_6$ | 110 | 011 | $s_3$ |
| $s_7$ | 111 | 011 | $s_3$ |

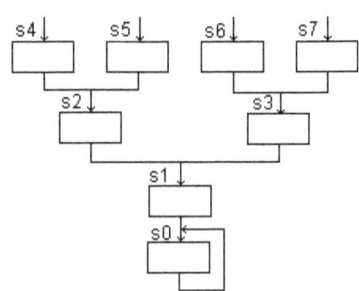

## 714 The ASM chart has 5 'loops'. 01281.., 26C93.., 5A5.., 7EDB7.., FF...

| PS | $q_3q_2q_1q_0$ | $q_3^+q_2^+q_1^+q_0^+$ | NS | PS | $q_3q_2q_1q_0$ | $q_3^+q_2^+q_1^+q_0^+$ | NS |
|---|---|---|---|---|---|---|---|
| $s_0$ | 0000 | 0001 | $s_0$ | $s_8$ | 1000 | 0001 | $s_1$ |
| $s_1$ | 0001 | 0010 | $s_2$ | $s_9$ | 1001 | 0011 | $s_3$ |
| $s_2$ | 0010 | 0100 | $s_4$ | $s_A$ | 1010 | 0101 | $s_5$ |
| $s_3$ | 0011 | 0110 | $s_6$ | $s_B$ | 1011 | 0111 | $s_7$ |
| $s_4$ | 0100 | 1000 | $s_8$ | $s_C$ | 1100 | 1001 | $s_9$ |
| $s_5$ | 0101 | 1010 | $s_A$ | $s_D$ | 1101 | 1011 | $s_B$ |
| $s_6$ | 0110 | 1100 | $s_C$ | $s_E$ | 1110 | 1101 | $s_D$ |
| $s_7$ | 0111 | 1110 | $s_E$ | $s_F$ | 1111 | 1111 | $s_F$ |

**715** The sequence of states is 701241...

| PS | $q_2q_1q_0$ | $q_2^+q_1^+q_0^+$ | NS |
|----|----|----|----|
| $s_0$ | 000 | 001 | $s_1$ |
| $s_1$ | 001 | 010 | $s_2$ |
| $s_2$ | 010 | 100 | $s_4$ |
| $s_3$ | 011 | 000 | $s_0$ |
| $s_4$ | 100 | 001 | $s_1$ |
| $s_5$ | 101 | 000 | $s_0$ |
| $s_6$ | 110 | 000 | $s_0$ |
| $s_7$ | 111 | 000 | $s_0$ |

**716** The sequence of states is 12436751243... etc. or it is locked in state 0/

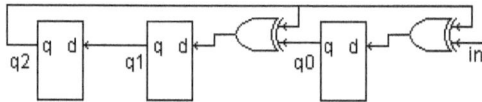

| $q_2q_1q_0$ | $d_2=q_1$ | $d_1=q_2\oplus q_0$ | $d_0=q_2\oplus in$ | $q_2^+q_1^+q_0^+$ |
|----|----|----|----|----|
| 000 | 0 | 0 | 0 | 000 |
| 001 | 0 | 1 | 0 | 010 |
| 010 | 1 | 0 | 0 | 100 |
| 011 | 1 | 1 | 0 | 110 |
| 100 | 0 | 1 | 1 | 011 |
| 101 | 0 | 0 | 1 | 001 |
| 110 | 1 | 1 | 1 | 111 |
| 111 | 1 | 0 | 1 | 101 |

**718** Add the divide by 12 reset circuit to Figure 718.

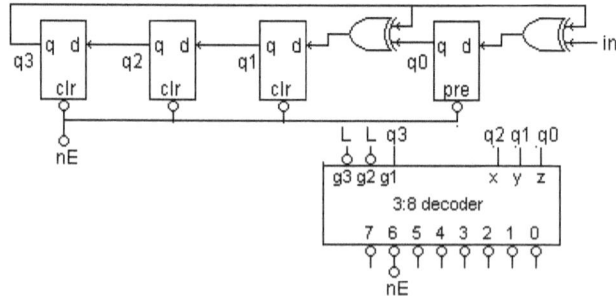

**717** The sequence of states is 1248 36CB 5A7E FD91... etc.

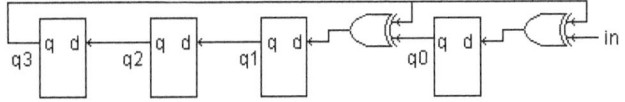

| $q_3q_2q_1q_0$ | $d_3=q_2$ | $d_2=q_1$ | $d_1=q_3\oplus q_0$ | $d_0=q_3\oplus in$ | $q_2^+q_1^+q_0^+$ |
|---|---|---|---|---|---|
| 0000 | 0 | 0 | 0 | 0 (in=0) | 0000 |
| 0001 | 0 | 0 | 1 | 0 | 0010 |
| 0010 | 0 | 1 | 0 | 0 | 0100 |
| 0011 | 0 | 1 | 1 | 0 | 0110 |
| 0100 | 1 | 0 | 0 | 0 | 1000 |
| 0101 | 1 | 0 | 1 | 0 | 1010 |
| 0110 | 1 | 1 | 0 | 0 | 1100 |
| 0111 | 1 | 1 | 1 | 0 | 1110 |
| 1000 | 0 | 0 | 1 | 1 | 0011 |
| 1001 | 0 | 0 | 0 | 1 | 0001 |
| 1010 | 0 | 1 | 1 | 1 | 0111 |
| 1011 | 0 | 1 | 0 | 1 | 0101 |
| 1100 | 1 | 0 | 1 | 1 | 1011 |
| 1101 | 1 | 0 | 0 | 1 | 1001 |
| 1110 | 1 | 1 | 1 | 1 | 1111 |
| 1111 | 1 | 1 | 0 | 1 | 1101 |

**719** Reference Figure 716 ($x^3+x+1$). Initial LFSR state is 000. Input bit stream is 10011010 ($x^8+x^5+x^4+x^2+1$), Show that sequence of states is 001, 010, 100, 010, 101, 001, 011, 110, 110, where 110 is the remainder. Also show that the $q_2$ sequence is 00101001 (the quotient).

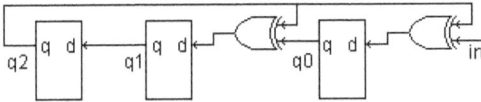

| $q_2q_1q_0$ | $d_2=q_1$ | $d_1=q_2\oplus q_0$ | in | $d_0=q_2\oplus in$ | $q_2^+q_1^+q_0^+$ |
|---|---|---|---|---|---|
| 000 | 0 | 0 | 1 | 1 | 001 |
| 001 | 0 | 1 | 0 | 0 | 010 |
| 010 | 1 | 0 | 0 | 0 | 100 |
| 100 | 0 | 1 | 1 | 0 | 010 |
| 010 | 1 | 0 | 1 | 1 | 101 |
| 101 | 0 | 0 | 0 | 1 | 001 |
| 001 | 0 | 1 | 1 | 1 | 011 |
| 011 | 1 | 1 | 0 | 0 | 110 |

Digital Design

**723** This is a very important problem. Convert Figures 720, 722, 723 into xyz bus format as shown in Figure 11 page 15.

Figure 720 converted to xyz bus format.

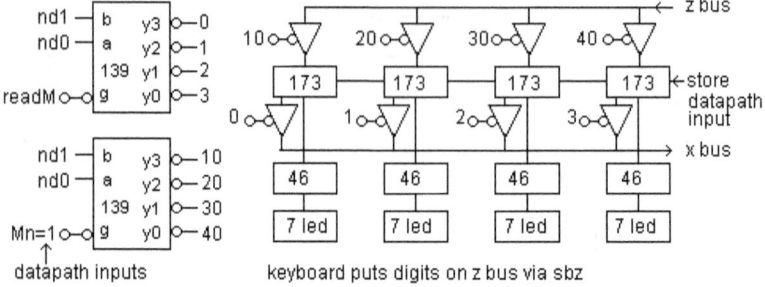

Figure 722 converted to xyz bus format.

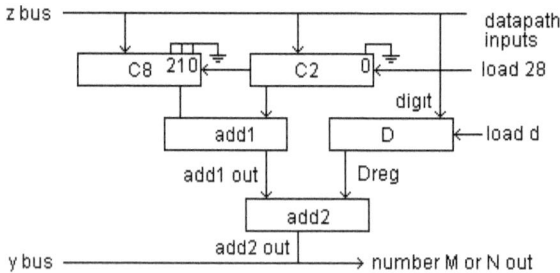

Figure 723 converted to xyz bus format.

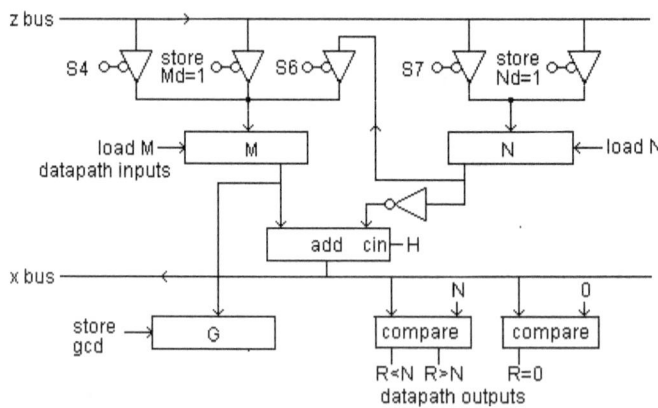

**802** For block 3, bank 5, $A_{19}$ thru $A_{14} = 011\ 101$. A13 thru A00 range from all 0 to all 1. Assembling the pieces we get 0000 00 011 101 xxxx xxxx xxxx xxxx xxxx. Regrouping we get 0000 0001 1101 xxxx xxxx xxxx xxxx xxxx or $01D00000_{16}$ to $01D0FFFF_{16}$

**803** Address lines are $A_F$ to $A_0$. Decoder spans the address space with 8K ($2^{13}$) banks.

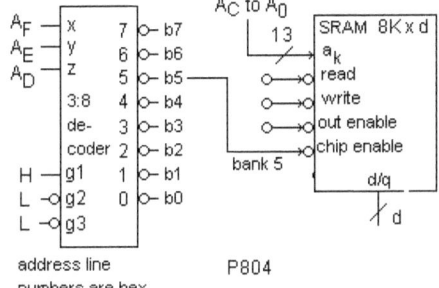

**804** Hex addresses 7BE0, 7BE1, 7BE2, 7BE3 in binary are 0111 1011 1110 00xx.

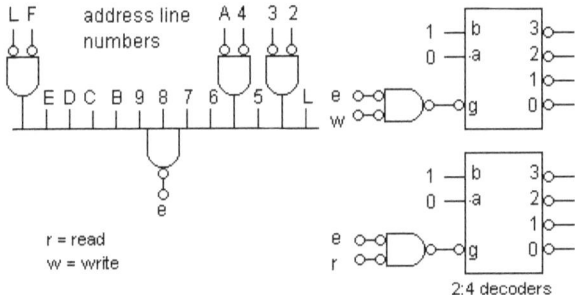

**805** Table shows bank RAM size is 64K, because sixteen address lines are required.

| Block address $A_{17}A_{16}A_{15}\ A_{14}$ | Bank address $A_{13}A_{12}A_{11}\ A_{10}$ | Chip address $A_{0F}$ to $A_{00}$ |
|---|---|---|
| $1010_2$ | $0000_2$ | 0000 to $FFFF_{16}$ |
| $1011_2$ | $0000_2$ | 0000 to $FFFF_{16}$ |
| $1100_2$ | $1000_2$ | 0000 to $FFFF_{16}$ |
| $1101_2$ | $1000_2$ | 0000 to $FFFF_{16}$ |

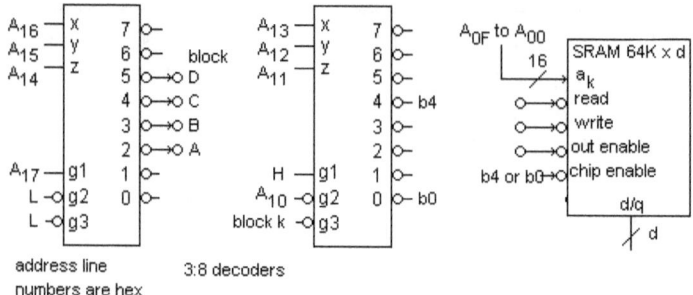

**806** The table shows $A_{1F}$ to $A_{18}$ lines are constant (do not change). AND these lines (133 and 02, not shown) to enable the block 138. Use constant $A_{17}$ to enable the block 138, which is addressed by $A_{16}A_{15}A_{14}$. Block 138 outputs A, B, C, D enable four bank 138s addressed by $A_{16}A_{15}A_{14}$. Each bank 138 output (b2, b3, b4, b5) enables a ram. RAM size is 64K. There are 16 rams.

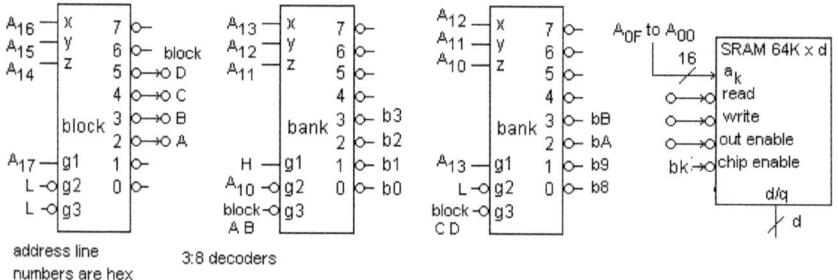

**808**

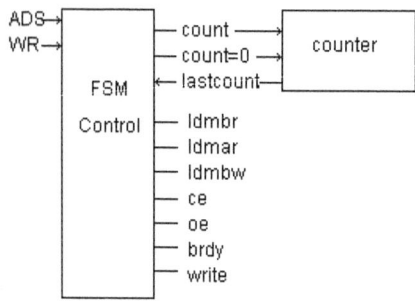

256

**809** State machine equations

$$q_0^+ = s_0 \cdot ads \cdot WR' + s_1 \cdot ltct'$$

$$q_1^+ = s_0 \cdot ads \cdot WR + s_2 \cdot ltct'$$

Output circuit equations

$$ce = s_0 \cdot ads + s_1 + s_2 \qquad oe = s_1$$

$$wr = s_0 \cdot ads \cdot WR + s_2 \qquad ct = (s_1 + s_2)ltct'$$

$$(ct = 0) = ldmar = s_0 \cdot ads$$

$$ldmbw = s_0 \cdot ads \cdot WR + s_2 \cdot ltct'$$

$$ldmbr = s_1 \cdot ltct$$

$$brdy = (s_1 + s_2)ltct$$

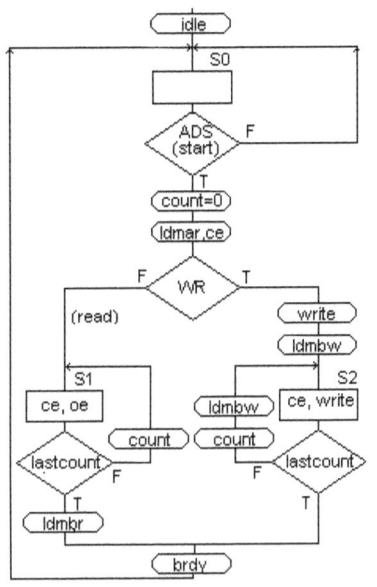

| $q_1q_0$ | ads | wr | ltct | $q_1^+q_0^+$ |
|---|---|---|---|---|
| 00 | 0 | - | - | 00 |
| 00 | 1 | 0 | - | 01 |
| 00 | 1 | 1 | - | 10 |
| 01 | - | - | 0 | 01 |
| 01 | - | - | 1 | 00 |
| 10 | - | - | 0 | 10 |
| 10 | - | - | 1 | 00 |

| $q_1q_0$ | ltct | ct=0 ldmar | ce | oe | ct | wr | ldmbw | ldmbr | brdy |
|---|---|---|---|---|---|---|---|---|---|
| 00 | - | 0 | 0 | 0 | 0 | 0 | 0 | 0 | 0 |
| 00 | - | 1 | 1 | 0 | 0 | 0 | 0 | 0 | 0 |
| 00 | - | 1 | 1 | 0 | 0 | 1 | 1 | 0 | 0 |
| 01 | 0 | 0 | 1 | 1 | 1 | 0 | 0 | 0 | 0 |
| 01 | 1 | 0 | 1 | 1 | 0 | 0 | 0 | 1 | 1 |
| 10 | 0 | 0 | 1 | 0 | 1 | 1 | 1 | 0 | 0 |
| 10 | 1 | 0 | 1 | 0 | 0 | 1 | 0 | 0 | 1 |

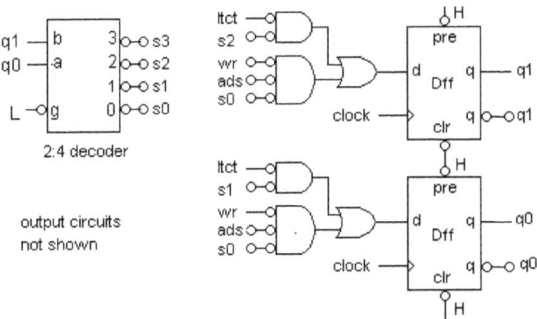

257

**810** A cycle by cycle description of a read miss cycle. Assume the ASM is idling in $S_0$, and *refresh* is F (Figure 816). Conditional output *count=0* is asserted. When *ads=1* AND *busy=0* AND write is F, the conditional output *read d* (read the dram) is immediately asserted. The read d command starts a dram read cycle. On the next clock edge the FSM steps from state $S_0$ to state $S_1$, and the SRAM starts a read cycle

While the ASM dwells in $S_1$, the sram and dram are accessing the data, and *count* is incremented on each clock edge. The *c done* report goes true in period 5 when sram access time *tacc* expires (this is defined by count = 3), and valid *data* becomes available. By this time the *match* is false. This is a miss that requires the dram read. On the next clock edge the FSM steps from state $S_1$ to state $S_3$ that allows the dram read to continue. When the dram read completes the *d done* report gores true asserting write c which write the new word from the dram into the cache sram, and *brdy* is asserted. On the next clock edge the FSM steps from state $S_3$ to state $S_0$.

**811** A cycle by cycle description of a write miss cycle. The ASM is in $S_0$ (Figure 816), and *refresh* is F. Conditional output *count=0* is asserted. When *ads=1* AND *busy=0* AND write is T, the conditional output *write d* is immediately asserted. *Count=0* resets the system logic access time timer to zero. *Write d* starts a dram write cycle regardless of a hit or miss in order to always have correct words in the main memory. On the next clock edge the ASM steps from state $S_0$ to state $S_2$. While the FSM dwells in $S_2$, the ASM is accessing the SRAM and DRAM data, and *count* is incremented on each clock edge. The *c done* report goes true in period 5 when the access time *tacc* expires (count equals 3). By this time *match* is false and this is a miss. A miss means the word being written into the dram does not have to be written into the cache sram (*write c* is not executed). At this point *brdy* is asserted. On the next clock edge the FSM steps from state $S_2$ to state $S_0$.

**1001** Add $200_{10}$ and $183_{10}$. What are NCZV values?

$200_{10} = 11001000_2 = C8_{16}$    $183_{10} = 10110111_2 = B7_{16}$

$11001000$

$\underline{+10110111}$

$101111111$   $\rightarrow$   $N = 0$   $C = 1$   $Z = 0$   $V = 1$

**1002** x=$200_{10}$, y=$183_{10}$. Do unsigned compare. What are NCZV values?

$200_{10} = 11001000_2 = C8_{16}$    $183_{10} = 10110111_2 = B7_{16}$

$11001000 - 10110111 = 11001000 + 01001000 + 1 = 100010001$

$\rightarrow$   $N = 0$   $C = 1$   $Z = 0$   $V = 1$

**1003** x=$200_{10}$, y=$183_{10}$. Do arithmetic (signed) compare. What are NCZV values?

$200_{10} = 11001000_2 = C8_{16}$    $183_{10} = 10110111_2 = B7_{16}$

$11001000 - 10110111 = 01111111$

$\rightarrow$   $N = 0$   $C = 0$   $Z = 0$   $V = 0$

**1004** Deduce the 8 bit word for the uI *LOAD 2B $r_A$*.

LOAD 2B $r_A$

| opcode | addr mode | reg | operation |
|--------|-----------|-----|-----------|
| load   | immediate | rA  | -         |
| 00     | 01        | 0   | 000       |

$uI = 00010000_2 = 10_{16}$

$iw = 00101011_2 = 2B_{16}$

**1005** Deduce the 8 bit word for the uI *STORE FF $r_A$*.

STORE FF $r_A$

| opcode | addr mode | reg | operation |
|--------|-----------|-----|-----------|
| store  | direct    | rA  | -         |
| 01     | 10        | 0   | 000       |

$uI = 01100000_2 = 60_{16}$

$iw = 11111111_2 = FF_{16}$

**1006** Deduce the 8 bit word for the uI *SUB A8 $r_A$*.

SUB A8 $r_A$

| opcode | addr mode | reg | operation |
|--------|-----------|-----|-----------|
| alu | immediate | rA | - |
| 10 | 01 | 0 | 001 |

$uI = 10010001_2 = 91_{16}$

$iw = 10101000_2 = A8_{16}$

**1007** Deduce the 8 bit word for the uI *BRne* –5.

BRne -5

| opcode | addr mode | reg | operation |
|--------|-----------|-----|-----------|
| Br | immediate | rA | BRne |
| 11 | 01 | 0 | 110 |

$uI = 11010110_2 = D6_{16}$

$iw = 11111011_2 = FB_{16} = -00000101_2 = -5_{10}$

**1008** Deduce the 8 bit word for the uI *XOR $r_A$ *$r_X(7)$*.

XOR $r_A$ *$r_X(7)$

| opcode | addr mode | reg | operation |
|--------|-----------|-----|-----------|
| alu | index | rA | XOR |
| 10 | 11 | 0 | 111 |

$uI = 10110111_2 = B7_{16}$

$iw = 00000111_2 = 07_{16}$

**1009** Write the mI list for the uI *XOR *$r_X(7)$ $r_A$*.

| $\underline{n}$ | *xor uI (index mode)* |
|-----|-----------|
| $n_0$ | $mar \leftarrow pc, \; pc \leftarrow pc + 1$ |
| $n_1$ | $ir \leftarrow M(mar), \; r$ |
| $n_2$ | $mar \leftarrow pc, \; pc \leftarrow pc + 1$ |
| $n_3$ | $r_T \leftarrow M(mar), \; r$ |
| $n_4$ | $mar \leftarrow r_X + r_T$ |
| $n_5$ | $r_T \leftarrow M(mar), \; r$ |
| $n_6$ | $r_A \leftarrow r_A \; xor \; r_T, \; n \leftarrow 0$ |

**1010** Write the mI list for the uI *BRun* $*r_X(7)$.

| $n$ | *brun uI* (*index mode*) |
|---|---|
| $n_0$ | $mar \leftarrow pc, \ pc \leftarrow pc + 1$ |
| $n_1$ | $ir \leftarrow M(mar), \ r$ |
| $n_2$ | $mar \leftarrow pc, \ pc \leftarrow pc + 1$ |
| $n_3$ | $r_T \leftarrow M(mar), \ r \ (store\,7 \ in \ r_T)$ |
| $n_4$ | $mar \leftarrow r_X + r_T$ |
| $n_5$ | $n \leftarrow 0, \ if \ mcc = 1, then$ |
| | $pc \leftarrow M(mar), \ r$ |

**1011** Write the logic equation for user data path input *srty*. Pretend many other registers can be connected to the y bus.

$$srty = n_4[load \cdot indx + store \cdot indx + alu \cdot (imm + indx) + br \cdot indx]$$
$$+ n_5[alu \cdot dir] + n_6[alu \cdot indx]$$

**1012** Write the logic equation for user data path input *sqz*.

$$sqz = \ n_1[load \cdot inh' + store(dir + indx) + alu \qquad + br \cdot inh']$$
$$+ n_3[load \cdot inh' + store(dir + indx) + alu \cdot inh' + br \cdot inh']$$
$$+ n_4[load \cdot dir \ + \qquad\qquad\quad + alu \cdot dir \ + br \cdot dir]$$
$$+ n_5[load \cdot indx + \qquad\qquad\quad + alu \cdot indx + br \cdot indx]$$

**1013** Write the logic equation for user data path input *spcz*.

$$spcz = n_0[load \cdot inh' + store(dir + indx) + alu \qquad + br \cdot inh']$$
$$+ n_2[load \cdot inh' + store(dir + indx) + alu \cdot inh' + br \cdot inh']$$

**1014** Write the logic equation for user data path input *zeron*.

$$zeron = \ n_2[ \qquad\qquad\qquad\quad + alu \cdot inh \qquad\qquad ]$$
$$+ n_3[load \cdot imm \qquad\qquad\qquad\quad + br \cdot imm]$$
$$+ n_4[load \cdot dir \ + store \cdot dir \ + alu \cdot imm + br \cdot dir \ ]$$
$$+ n_5[load \cdot indx + store \cdot indx + alu \cdot dir \ + br \cdot indx]$$
$$+ n_6[ \qquad\qquad\qquad\qquad + alu \cdot indx \qquad\quad ]$$

**1015** Write the logic equation for user data path input read $r$. It is the same a *smemz*.

$$r = n_1[load \cdot inh' + store(dir + indx) + alu \qquad + br \cdot inh']$$
$$+ n_3[load \cdot inh' + store(dir + indx) + alu \cdot inh' \quad + br \cdot inh']$$
$$+ n_4[load \cdot dir \ + \qquad\qquad\quad + alu \cdot dir \quad + br \cdot dir]$$
$$+ n_5[load \cdot indx + \qquad\qquad\quad + alu \cdot indx + br \cdot indx]$$

**1016** Write the logic equation for user data path input *ldst*.

$$ldst = n_2[alu \cdot inh] + n_4[alu \cdot imm] + n_5[alu \cdot dir] + n_6[alu \cdot indx]$$

**1017** Write the logic equation for user data path input *ldmar*.

$$ldmar = n_0[load \cdot inh' \ + store(dir + indx) + alu \qquad + br \cdot inh']$$
$$+ n_2[load \cdot inh' \ + store(dir + indx) + alu \cdot inh' \ + br \cdot inh']$$
$$+ n_3[load \cdot dir \ + store \cdot dir \qquad\quad + alu \cdot dir \ + br \cdot dir]$$
$$+ n_4[load \cdot indx + store \cdot indx \qquad + alu \cdot indx + br \cdot indx]$$

**1018** Write the logic equation for user data path input *ldra*. Pretend many other registers can be connected to the x bus.

$$ldra = n_2[ \qquad\qquad + alu \cdot inh \ ]$$
$$+ n_3[load \cdot imm \qquad\qquad ]$$
$$+ n_4[load \cdot dir \ + alu \cdot imm]$$
$$+ n_5[load \cdot indx + alu \cdot dir \ ]$$
$$+ n_6[ \qquad\qquad + alu \cdot indx]$$

**1019** Write the logic equation for user data path input *saz*

$$saz = \ n_2[ \qquad\qquad\qquad + alu \cdot inh \qquad\qquad ]$$
$$+ n_4[load \cdot indx + store \cdot indx + alu(imm + indx) + br \cdot indx]$$
$$+ n_5[ \qquad\qquad\qquad + alu \cdot dir \qquad\qquad ]$$
$$+ n_6[ \qquad\qquad\qquad + alu \cdot indx \qquad\qquad ]$$

# Fourteen Experiments

The experiments support the text. You teach yourself how to use electronic instruments and tools, how to make measurements, and learn about parts. You teach yourself how to build the circuits shown in the figures on a solderless breadboard, and measure their performance.

Electrical engineering book learning is necessary but not sufficient, because an EE has to know

1 how to make measurements in order to evaluate a design.
2 what parts are available as well as their properties.
3 what parts look like, and how to read any part's label.
4 the equivalent circuit of a part given a frequency range,
5 that all parts have parasitic components attached to them.
6 what parasitic components are added when a part is placed in a circuit.
7 how to use the tools of the trade
8 and so on.

Here you design and build circuits from the get go, while referring to the text and other writings to learn what you need to know to implement the design. The idea is that you seek answers on an ongoing basis to the many questions that arise as you try to design and build circuits.

You take time outs to seek answers to the questions by reading the text, doing the problems, and perhaps doing a search on the Internet. In this way theory and practice merge.

## Parts

A comprehensive view of available parts is found in the Product Index on the web site of any electronics parts distributer. The experiments only use a small subset of each type of part such as resistors R, capacitors C, inductors L, transformers, potentiometers R, transistors, and integrated circuits.

Associated with every part is a *data sheet*, which presents the part's characteristics such as text descriptions, thermal characteristics, tables of electrical characteristics, available electrical values, package dimensions, pin assignments, test circuits, application notes, and so forth.

A *data sheet* tells you what a part is about.

Rarely will you find in a *data sheet* the equivalent circuit of a part given a frequency range, nor any information about the part's parasitic components attached to them. This information is usually found in *technical articles* or *application notes* issued by the manufacturer.

Every manufacturer has a web site from which you can download data sheets, Spice models, application notes, white papers and so forth.

*Each experiment specifies the parts used in the experiment's circuits.*

# Instruments

What do you need to measure or observe, while evaluating circuit performance? As a minimum you need to measure or observe ohms, DC voltages, DC currents, steady state AC signal voltages, and transient state signal voltages.

You need an oscilloscope so that you can see the DC voltages and AC signal voltages at circuit nodes. Two channels allow you to compare what is going on at two nodes, such as an input and its corresponding output. For our purposes here a two channel oscilloscope, with 1MHz bandwidth or better, is satisfactory.

You need a signal generator, specifically a function generator, to generate the signals driving the circuit under test. Function generator, because you will need sine waves, square waves, and pulses. A function generator with maximum frequency 1MHz or better will do.

*A very important feature of any signal generator is that a signal amplitude does NOT change when frequency is changed. This must be verified as experiments are executed.*

You need a DC multimeter, which measures wide ranges of volts, ohms, and amperes. Analog or digital meter? Your choice. We use both types.

To us a Power Supply is a *generator* of zero frequency DC voltages. In this text's experiments you need a ±5V linear power supply.

# Tools

Side cutters and long nose pliers cut and form wire. A wire stripper strips insulation from the ends of wire without damaging the wire. A lead forming tool accurately forms wire leads of parts for insertion into the breadboard holes. Tweezers facilitate picking up and placing small parts. Clip leads connect terminals as needed.

1   breadboard
1   IC extractor
1   IC Inserter
1   side cutters, 4
1   long nose pliers, 4
1   wire stripper
1   lead forming tool
1   tweezers, fine point
?   Clip leads

### Color Code

An effective way to learn the color code is to sort a pile of resistors by value. Another way is to use the program *colorcode.exe*.

| *digit* | 0 | 1 | 2 | 3 | 4 | 5 | 6 | 7 | 8 | 9 |
|---|---|---|---|---|---|---|---|---|---|---|
| *color* | black | brown | red | orange | yellow | green | blue | violet | gray | white |
| *10% values* | 100 | 120 | 150 | 180 | 220 | 270 | 330 | 390 | 470 | 560 | 680 | 820 | 1000 |

| *5% values* | | | | | | | | | | | | |
|---|---|---|---|---|---|---|---|---|---|---|---|---|
| 100 | 110 | 120 | 130 | 150 | 160 | 180 | 200 | 220 | 240 | 270 | 300 | 330 |
| 360 | 390 | 430 | 470 | 510 | 560 | 620 | 680 | 750 | 820 | 910 | 1000 | |

1st band    1st digit
2nd band   2nd digit
3rd band   number of zeros
4th band   tolerance, gold 5%, silver 10%

Examples
22      red, red, black
220K   red, red, yellow
1.2K    brown, red, red
47K    yellow, violet, orange
910    white, brown, brown
8.2M   gray, red, green

Digital Design

# Contents

# SAFETY FIRST!!!!!!!

SAFETY FIRST!!!!!!!! Electricity is silent so be very careful. We remove all metal objects from our hands and wrists such as rings and watches. If you do not remove them, then know you are taking an unnecessary risk.

Furthermore you do the experiments at your own risk, because there is no way we can supervise your work.

Word to the wise - NEVER GRAB anything, because if its hot you will have difficulty letting go.

**The AC Line Voltage is extremely dangerous.**

# The Solderless Breadboard and the Power Supply

Solderless breadboards are designed to connect parts together without using solder. A power supply energizes the circuits on the solderless breadboard.

You build a circuit by inserting leads in holes. For example a resistor has 2 leads, which when suitably bent can be inserted into 2 pin holes. There is no need to shorten a part's leads in these experiments.

A part's lead (28 to 20 AWG, 0.0126 to 0.0320 inches diameter) is inserted into a hole whose spring loaded metal insert grabs the lead. The metal inserts are not visible. *The metal inserts in horizontal rows of 5 holes are shorted together.* Consequently leads inserted in the same 5-hole-row are shorted together. The leads placed in a 5 hole row are the equivalent of leads soldered together. A row is equivalent to a circuit node.

**Figure 101 Part of a Solderless breadboard showing fields of holes.**

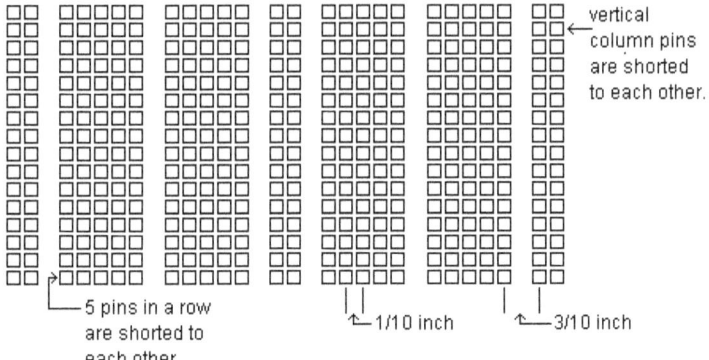

A board is an assembly of *two types of hole patterns* mounted on a metal plate (Figure 101). One type of hole pattern has two $5 \times 59$ arrays of holes separated by a narrow gutter. The columns of rows parallel and adjacent to the gutter are 0.3 inches apart, because 0.3 inches is the IC DIP package minimum pin row spacing. Each array of holes is a column of 59 five hole rows. All holes are on a 0.1 inch grid so that vertical and horizontal separation is 0.1 inch. The five holes in each row are shorted together. The rows are not shorted together.

IC pins are inserted into the board so that the IC straddles the gutter and each IC pin plugs into one hole of one row. Then the four other holes in each pin row are available to receive leads, which are automatically connected to the IC pin. In this way a circuit is wired from node to node. Larger ICs have 0.4-inch and 0.6-inch pin row spacing. These are inserted in the same way that the narrower 0.3 inch ICs are inserted; however, the covered up row holes are not available for point-to-point wiring.

The 3 leads of a discrete transistor are inserted in 3 *different* rows.

The 3 leads of a potentiometer are inserted in 3 *different* rows

***Power and Ground***   The other type of hole pattern is formatted to distribute power and ground. The pattern has two columns of 50 holes. The 50 holes *in each column* are shorted together. The two columns are not shorted together so that one column can distribute 5 volts, for example, and the other 0 volts (ground). Another pattern across the top of the board has two rows of 40 holes. The 40 holes *in each row* are shorted together.

Above the rows of 40 holes the solderless breadboard has binding posts whose insulated knobs unscrew to reveal a hole in the post in which a wire is inserted. The other end of the wire is plugged into a row of 40 holes (or a column of 50 holes). In turn jumpers connect the 40 hole rows to the 50 hole columns. (The black post is shorted to the metal base plate, and the red ones are insulated from the base plate.)

### *Verifying the solderless breadboard shorts*

Select the R × 10 or higher ohm range, because the R × 1 range drains the 1.5 volt battery (use the R × 1 range only when you have to).

Connect a pin to each lead of your ohmmeter. Place one pin in the first hole of a 5 hole row. Place the other pin in each of the other 4 holes in the row. Verify that the resistance is essentially zeros ohms (a short).

Repeat for the columns of holes.

Measure the resistance between rows, which should be infinite (open circuit). Repeat for row to columns, etc. Check out all possibilities to KNOW how the holes are wired.

*Power Supply* You can select any ±5V power supply >20 watts. We use a linear open frame ±5V power supply. We have to add a power cord. COVER THE AC TERMINALS WITH TAPE. Plug the power cord into a plug strip outlet. The plug strip on/off switch becomes the power supply on/off switch. Think of a power supply as a constantly recharged battery.

The source impedance $R_S$ of the battery is estimated as follows. If a 5V output drops by 2%, or 0.1V, when a 0.25 ampere load current is drawn, then $R_S$=0.1V/0.25A=0.4 ohms. However if plus is shorted to minus, then potentially I=5/0.4=12.5 amperes, which may or may not flow. This is why unprotected parts are destroyed (and perhaps the power supply).

We have to be very careful to avoid shorting plus to minus.

> *Shut off the power supply OR Disconnect the voltage leads AT THE SUPPLY* when making changes on the solderless breadboard.

### Connecting the power supply to the solderless breadboard

*Connect the binding posts to the solderless breadboard.* The solderless breadboard has binding posts whose insulated knobs unscrew to reveal a hole in the post. The black post is shorted to the metal base plate, and the red ones are insulated from the base plate. Unscrew the black post and insert a black wire in the post hole. Tighten the black knob to secure the wire. Insert the other end of the wire into a hole in a column of holes that you want to be grounded. Repeat with a red wire from a red post to what becomes the power supply voltage, the B+ column of holes (e.g. +5V). Repeat for B− (e.g. −5V).

*Connect the power supply to the binding posts* First turn off the AC power to the supply. Use clip leads or wires to connect the power supply voltage terminals to the solderless breadboard binding posts.

AFTER you have built a circuit, turn on the AC power to the supply.

Turn the power off *before* you make circuit changes.

Use the multimeter to verify the ±5V voltages.

Digital Design

# Experiment 1 NAND, NOR, NOT, XOR Gates

1 Build a circuit (Figure E101, E102, E103). Use wires to connect the voltage sequence LL, LH, HL, HH at the inputs and note which led lights up. I.e. replicate the truth table. Is green or red a 1 (on) or a 0 (off)?

Then connect one input to H and the other to the output of a square wave signal generator. Set the amplitude to 0/5 volt. Verify with an oscilloscope.

Set frequency to 1Hz or lower so that you can observe the Green-Red on-off sequence. Which truth table rows does this sequence represent?

2 Build a circuit (Figure E104). Connect the signal generator to the 163 clock input. Select 100KHz frequency. Sync scope to $R_{CO}$. Observe truth table output sequences. 00 HHHL, 02 LHHH, 86 LHHL.

**Figure E101 NAND 00**          **Figure E102 NOR 02**

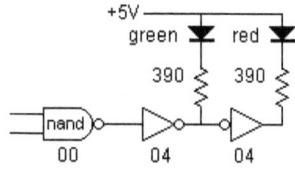

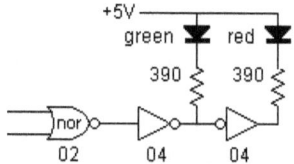

**Figure E103 XOR 86**

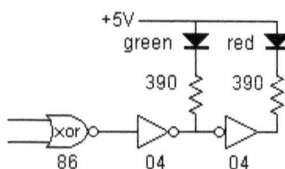

**Figure E104**

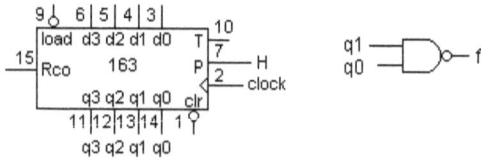

270

# Experiment 2 Multiplexers

## 1. Multiplexer Application: Function Generators

To be specific suppose we want to use a mux to create an xor gate. The XOR function's output column of four 1's and 0's would be 0110. The 0110 column aligns with truth table rows 0, 1, 2, 3. so we connect LHHL to the inputs of a 4 to 1 mux, and input variables yz to the mux ba address line inputs (Figure E201). The mux now stores the xor function.

**Figure E201 Generating the xor function**

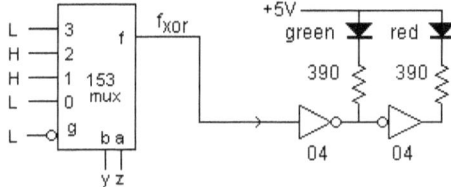

Build the circuit (Figure E201). Connect the voltage sequence LL, LH, HL, HH at the mux b, a inputs and note which led lights up. I.e. replicate the xor truth table. Is green, red a 1 (on) or a 0 (off)?

Then connect one input to H and the other to the output of a square wave signal generator. Set the amplitude to 0/5 volt. Verify with an oscilloscope. Sync the scope to $f_{xor}$.

Set frequency to 100KHz so that you can observe the LHHL sequence.

Digital Design

2. Multiplexers can implement a function with table entered variables. *The data inputs are the mux table entered variables.*

*jk function* Any mux can create the jk function $f_{jk}=jq'+k'q$.
$4:1\,mu\,x$

$$f_{4\_jk} = jq'+k'q = j(k'+k)q'+(j'+j)k'q$$
$$= q'(jk')+q'(jk)+q(j'k')+q(jk')$$
$$= q'm_2 + q'm_3 + qm_0 + qm_2$$
$$= q'\cdot m_3 + 1\cdot m_2 + 0\cdot m_1 + q\cdot m_0$$

**Figure E202 Generating the jk function**

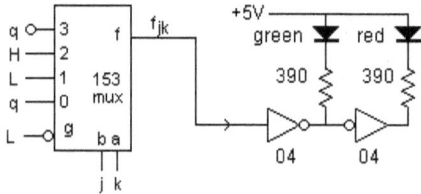

Build the circuit (Figure E202). Connect the voltage sequence LL, LH, HL, HH at the mux b, a inputs and note which led lights up. I.e. replicate the jk truth table.

Run the experiment when q=H, q'=L, and again when q=L, q'=H.

Then connect the b input to H and the a input to the output of a square wave signal generator. Set the amplitude to 0/5 volt. Verify with an oscilloscope.

Set frequency to 1Hz so that you can observe the Green-Red on-off sequence. Which truth table rows does this sequence represent?

Reverse the a, b inputs and repeat the experiment.

---

3 Design and build a 4-bit up/down counter with clear c, count p, load g , up/down u controls. Use 153, 163, and gates.

---

# Experiment 3 Decoders

1 Build the circuit in Figure E302. Connect a function generator output to the 163 clock input. Select a squarewave output waveform ranging from 0V to 5V. Set frequency to 100KHz (period = 10μs).

Connect oscilloscope channel 1 to the $y_0$ signal at the 139 output. Sync the scope to $y_0$. Select timebase 10μs/cm.

In turn connect channel 2 to $y_0$, $y_1$, $y_2$, $y_3$ to observe $y_k$ waveforms shown in Figure E301.

**Figure E301 Decoder Waveforms**

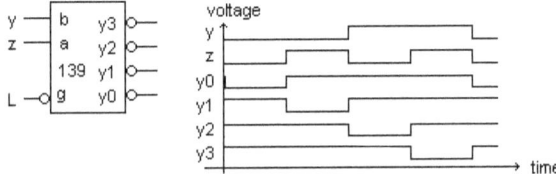

**Figure E302 Decoder Circuit**

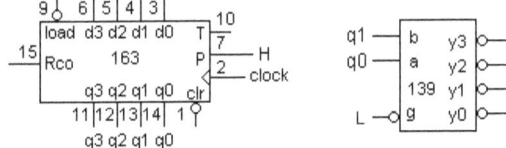

2 Design a counter that recycles as it counts from $07_{16}$ to $D4_{16}$. Use two 163, two 138, and gates.

Digital Design

# Experiment 4 Binary Arithmetic

Build the circuit in Figure E401.

Use 163 outputs $q_3$, $q_2$, $q_1$, $q_0$ to drive the $a_K$, $b_K$ inputs as shown in Figure E401. Connect $c_{in}$ to L.

Predict the sums and carries. Hint: draw the $q_3$, $q_2$, $q_1$, $q_0$ waveforms. Below them draw the sum waveforms (without carries), the carry out waveforms, and then the *sum* of sums and carry outs.

Connect a function generator output to the 163 clock input. Select a squarewave output waveform ranging from 0V to 5V. Set frequency to 100KHz (period = 10µs). Sync the scope to $R_{co}$.

At each node there is a 1 0 sequence representing the waveform. Compare the sequences to your predictions.

**Figure E401 4-bit binary Full adder 283**

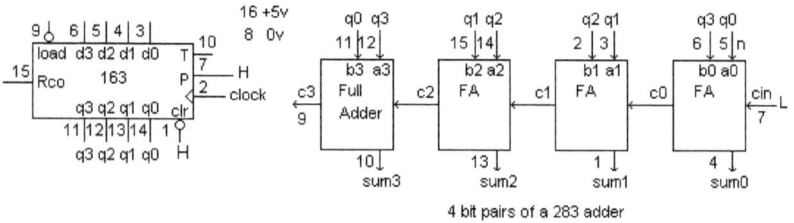

4 bit pairs of a 283 adder

# Experiment 5 Comparing Numbers

Build the 163, 85 circuit in figure E501.

Predict the a<b, a=b, and a>b outputs. See hint in experiment 4.

Connect a function generator output to the 163 clock input. Select a squarewave output waveform ranging from 0V to 5V. Set frequency to 100KHz (period = 10μs). Sync the scope to $R_{co}$.

At each node there is a 1 0 sequence representing the waveform. Compare the sequences to your predictions.

**Figure E501 4 Bit Comparator 85**

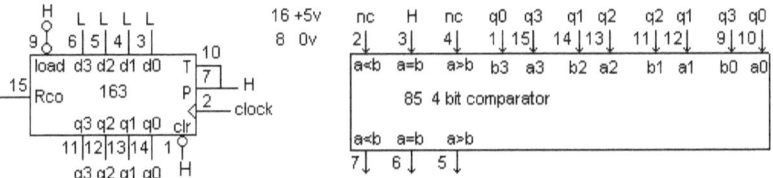

Digital Design

# Experiment 6 Flip-flops

Connect a function generator output to the clock inputs. Select a
squarewave output waveform ranging from 0V to 5V. Set frequency to
100KHz (period=10μs). Sync the scope to the right hand output waveform
$q_1$ or $q_2$. At each d and q node there is a 1 0 sequence representing the
waveform. Compare the sequences to your predictions.

Build the circuit in Figure E601. Starting with all flip-flop outputs at the
q=L level predict $q_0$, $q_1$ over five clock periods.

**Figure E601**

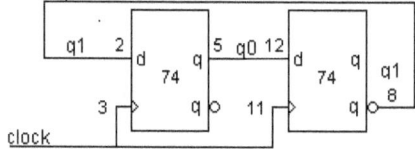

Build the circuit in Figure E602. Starting with all flip-flop outputs at the H
level predict $q_0$, $d_1$, $q_1$ over six clock periods.

**Figure E602**

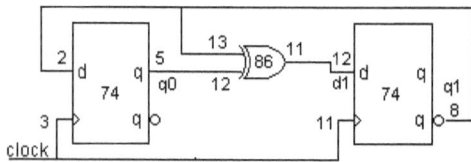

Build the circuit in Figure E603. Starting with all flip-flop outputs at the H
level predict $q_0$, $d_1$, $q_1$, $d_2$, $q_2$ over nine clock periods. Note: if there is no
waveform turn power off, wait several seconds, then turn power on.

**Figure E603**

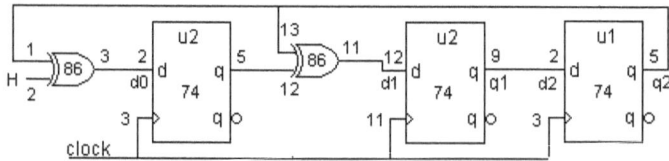

# Experiment 7 Counters

Sync the scope on n$_J$. Are your waveforms the same as in the Figures?

1 Build the Problem 710 waveform generator circuit. Do not copy the solutions in Answers to some of the Problems. Design the circuit from the waveforms.

**Figure P710a Digital Waveforms**

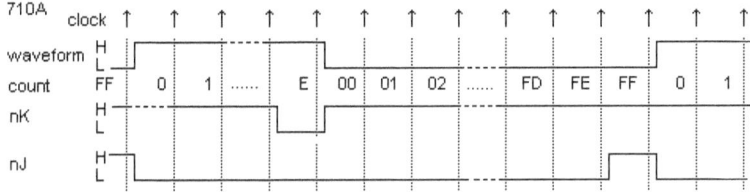

2 Build the Problem 711 waveform generator circuit. Do not copy the solution. Design the circuit from the waveforms.

**Figure P711a Digital Waveforms**

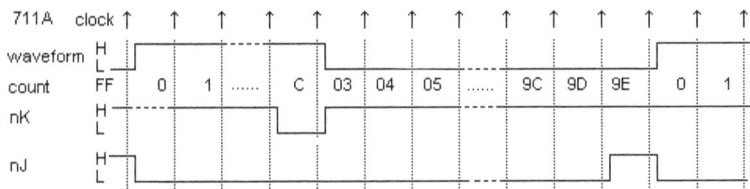

# Experiment 8 Shift Registers

Design a 4 bit or 8 bit general purpose shift register that can perform the operations in the table below.

Use 2 each 175 quad $d_{FF}$, six or ten each 153 4:1 mux, 1 each 04 as out buffers

Programming:
$s_5s_4$ select an input ($q_0$ in $q_7$) from the left
$s_3s_2$ select data $d_K$, q from the left, q from the right, (hold) $q_K$
$s_1s_0$ select an input ($q_7$ in L) from the right

Hints:
What are the four possible $d_{FF}$ d inputs? Answer data, qleft, qright, q
What is different about the q7 $d_{FF}$ and the q0 $d_{FF}$?
How many bits program a 153?

Draw a schematic.

**Figure E801**

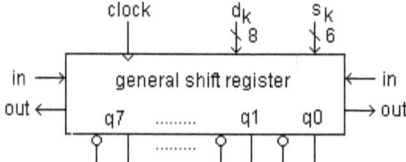

| Operator | Bit pattern | | Program bits | |
|---|---|---|---|---|
| *Before* | $q_7\ q_6\ q_5\ q_4\ q_3\ q_2\ q_1\ q_0$ | | $s_5\ s_4\ s_3\ s_2\ s_1\ s_0$ | |
| | *After one shift* | | *– is don't care* | |
| ROTR | $q_0\ q_7\ q_6\ q_5\ q_4\ q_3\ q_2\ q_1$ | | 0 0  1 0  – – | |
| ROTL | $q_6\ q_5\ q_4\ q_3\ q_2\ q_1\ q_0\ q_7$ | | – –  0 1  0 0 | |
| SR | in $q_7\ q_6\ q_5\ q_4\ q_3\ q_2\ q_1$ | | 0 1  1 0  – – | |
| SL | $q_6\ q_5\ q_4\ q_3\ q_2\ q_1\ q_0$ in | | – –  0 1  0 1 | |
| SRA | $q_7\ q_7\ q_6\ q_5\ q_4\ q_3\ q_2\ q_1$ | | 1 0  1 0  – – | |
| SLA | $q_6\ q_5\ q_4\ q_3\ q_2\ q_1\ q_0$ 0 | | – –  0 1  1 0 | |
| LOAD | $d_7\ d_6\ d_5\ d_4\ d_3\ d_2\ d_1\ d_0$ | | – –  1 1  – – | |
| HOLD | $q_7\ q_6\ q_5\ q_4\ q_3\ q_2\ q_1\ q_0$ | | – –  0 0  – – | |

A solution is available in SOLUTIONS.

# Experiment 9 Single Pulse, Debounced, Synchronized

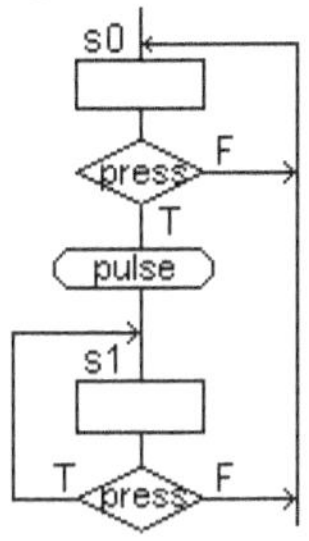

A SPDT switch (toggle/push button) has a single pole double throw format required by a switch debouncer. The switch is in the r position (Figure E902). An asynchronous switch *press* (move) to s sets the circuit so that q is H. A debounced press as input to a synchronizer becomes a synchronized *press* (Figure E903).

**Figure E901**

```
         s0  ◄─────────┐
         ┌──┐          │
         └──┘          │
     ┌─────────┐  F    │
     ◄ press ──────────►
     └─────────┘       │
          T            │
     ( pulse )         │
        │              │
         s1  ◄─────┐   │
         ┌──┐      │   │
         └──┘      │   │
      T            │   │
     ◄ press ──────┘   F
     └─────────┘───────►
```

Create a truth table and equations for the circuit. Design a circuit implementing the ASM (Figure E901) that emits a pulse of one system clock duration when press is pressed. The ASM is clocked by the system clock. Use a flip-flop output as the pulse source, because commercial flip-flops are designed to be free of hazards. Draw a schematic. A solution is available in SOLUTIONS.

As for Debounced consider the following. The arm of a switch departing from one position creates a series of break and make connections at this departing contact, travels for a finite time connected to nothing, and makes a series of make and break connections at the arriving contact. A switch debouncer filters out all of this "noise" and produces clean changes of state at its outputs q and z (Figure E902).

**Figure E902 Switch Debouncer**

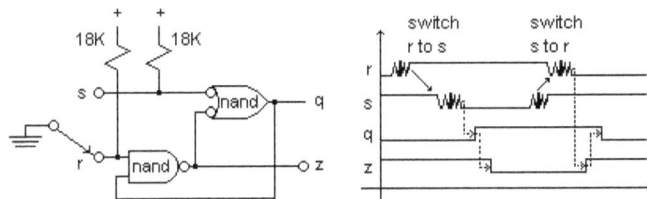

**Figure E903 One Way to Synchronize**

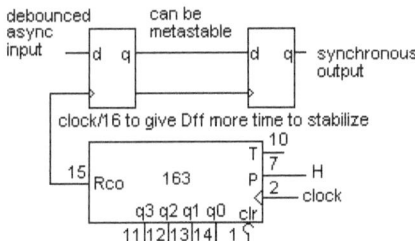

# Experiment 10 System Clock

A useful system clock has two modes: continuous clock *sysclk* and single pulse.

The two state ASM (Figure E1001a) divides the system clock by two producing *sysclk*. In Figure E1001b when input *clk* goes F *sysclk* stops at state $S_0$.

In Figure E1001c when input *clk* goes F *sysclk* stops at H or L depending on which state is the last state. Since *press* is F the system ASM immediately moves to $S_0$ where *sysclk* is L.

See Experiment 9 for *press*. Here *clk* is another input like *press*.

With *clk* at F (position r) moving the press switch from r to s emits *press*, which moves the ASM from $S_0$ to $S_1$ where it remains until the pres switch is switched back to r. Then the ASM moves to $S_0$.

Observe that the pulse is H until press is switched back to F (r) or input *clk* is switched back to T (s).

The *clk* and *press* inputs are debounced and synchronized.

Create a truth table and equations for the circuit. Design a circuit implementing the ASM (Figure E901). Draw a schematic.

**Figure E1001 Evolution of an ASM**

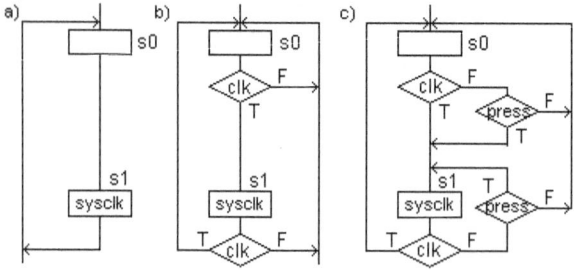

A solution is available in SOLUTIONS.

# Experiment 11 Receiver Serial Clock

A sender outputs a serial data stream using a clock frequency, *data clock*, known to the receiver. The receiver needs to create a local clock, *rclk*, so that the serial data stream *data d* can be converted into a 1,0 stream (Figure E1101)

**Figure E1101 Data Stream creates Receiver clock**

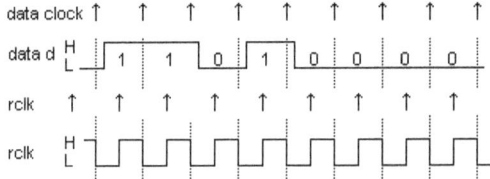

To detect a change in the serial data stream the receiver compares data values in successive clock periods using an XOR gate. The XOR gate, with output *xor*, produces an L to H, LH, transition at each 01 or 10 pair, and an H to L, HL, transition or no transition at 00 or 11 pairs.

Assume the data stream d is debounced $D_{FF\_1}$ with output *data d* (Figure E1101). Synchronize *data d* drives a $D_{FF\_2}$ to get a one clock delayed output. The $D_{FF\_2}$ delayed output is the *dprior* bit, and the $D_{FF\_1}$ *data d* output is the *dnew* bit. Let *xor= dprior $\oplus$ dnew*.

The receiver circuit includes a counter counting at 16 times data clock. The counter outputs *15* and *7* when count equals 15 and 7. The counter is reset to 0 on *xor* or *15*. The counter counts up on *xor′ × 15′*.

**Figure E1102**

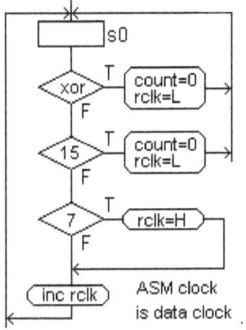

Each 10, 01 transition, where *xor = 1*, resets the counter. At 7 a $JK_{FF}$ is set, and at 15 or an LH the $JK_{FF}$ is reset. The $JK_{FF}$ output is *rclk*, which samples the serial data stream to produce the 1,0 stream.

Create a truth table and equations for the circuit. Design a circuit implementing the ASM (Figure E1102). Draw a schematic. How would you test the circuit? A solution is available in SOLUTIONS.

Digital Design

# Experiment 12 SRAM Read/Write ASM to circuit

Create a truth table with headings present state, ADS, WR, lastcount, next state such as Truth Table 703 page 150. Do not include the outputs.

The truth table represents a state machine. Derive the state machine's $D_{FF}$ equations from the truth table.

You can save time by not including the outputs in the truth table, because you can write the output equations by examining the ASM chart.

Design the state machine circuit including the output logic.

Draw a schematic.

A solution is available in SOLUTIONS.

**Figure 809 Read/Write ASM**

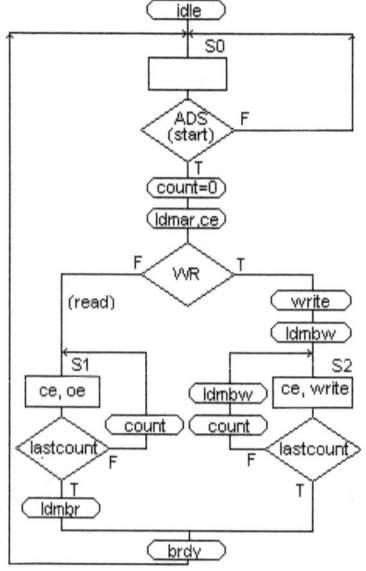

# Experiment 13 Cache ASM to circuit

Create a truth table such as Truth Table 703 page 150 or 101 on page 10. Do not include the outputs.

Derive the state machine's $D_{FF}$ equations from the truth table.

You can save time by not including the outputs in the truth table, because you can write the output equations by examining the ASM chart.

Design the state machine circuit including the output logic.

Draw a schematic.

A solution is available in SOLUTIONS.

**Figure 816 Cache ASM**

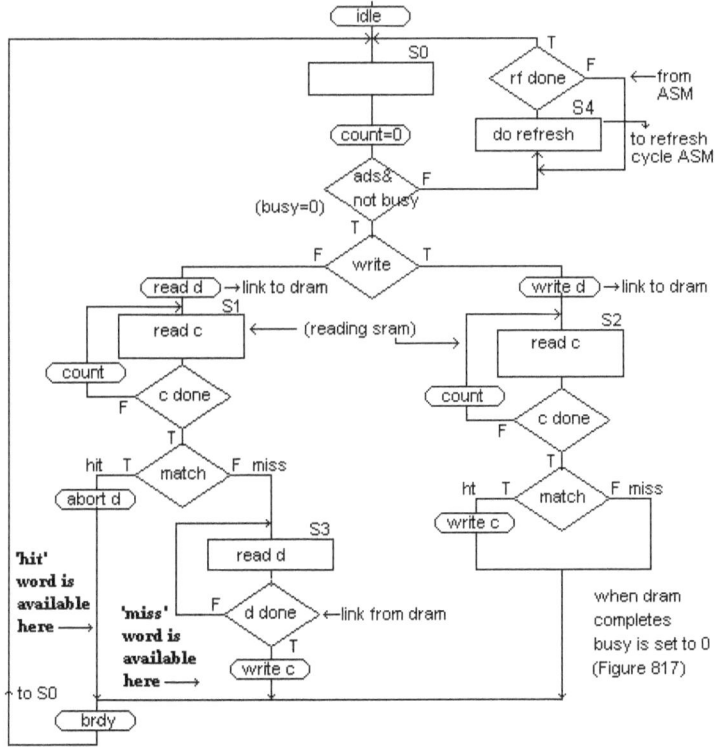

# Experiment 14 Hamming Code Encoder and Decoder

Use the Hamming G matrix to design an encoder that converts 4 message bits (k=4) into a 7 bit codeword.

Use the Hamming H matrix to design a decoder that converts a 7 bit codeword into a 3 bit syndrome S.

Draw schematics. Reference Figures 818, 819.

Here when we add and multiply 1's and 0's, the operations on the binary numbers are executed modulo 2 so that all results are 1 or 0 (1+1=0, 1×1=1, 1×0=0). See equations 1, 2, 3 page 177.

A solution is available in SOLUTIONS.

Here are the Hamming G and H matrices.

*encoder*  $C = \begin{bmatrix} c7 & c6 & c5 & c4 & c3 & c2 & c1 \end{bmatrix}$

$C = M \times G$

$$= \begin{bmatrix} k4 & k3 & k2 & k1 \end{bmatrix} \times \begin{bmatrix} 1 & 0 & 0 & 1 & 0 & 1 & 1 \\ 0 & 1 & 0 & 1 & 0 & 1 & 0 \\ 0 & 0 & 1 & 1 & 0 & 0 & 1 \\ 0 & 0 & 0 & 0 & 1 & 1 & 1 \end{bmatrix}$$

*decoder*  $S = HC^T$

$$H = \begin{bmatrix} 1 & 1 & 1 & 1 & 0 & 0 & 0 \\ 1 & 1 & 0 & 0 & 1 & 1 & 0 \\ 1 & 0 & 1 & 0 & 1 & 0 & 1 \end{bmatrix}$$

## SOLUTIONS
## Experiment 8 Figure E802

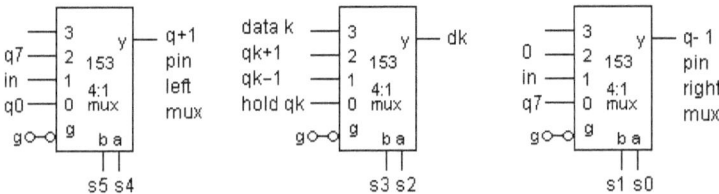

## Experiment 9

| ps $q_1q_0$ | press | ns $q_1^+q_0^+$ | Pulse |
|---|---|---|---|
| 0 | 0 | 0 | |
| 0 | 1 | 1 | 1 |
| 1 | 0 | 0 | |
| 1 | 1 | 1 | |

$$q_0^+ = press\ S_0 + press\ S_1 = press(q_0' + q_0) = press$$
$$pulse = press\ S_0 = q_0'\ press$$

Note: you can see the single pulse if 163 $q_3$ is the synchronous press and you sync the scope with + going $q_3$.

**Figure E904**

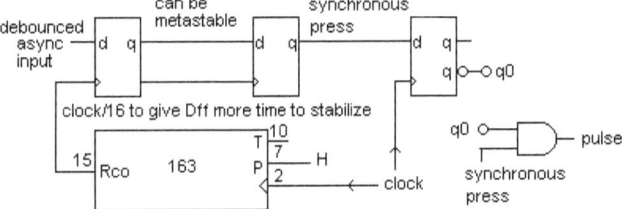

| | Experiment 10 |
|---|---|
| q | clock | press | $q^+$ |

Combined table:

| q | clock | press | $q^+$ |
|---|---|---|---|
| 0 | 0 | 0 | 0 |
| 0 | 0 | 1 | 1 |
| 0 | 1 | – | 1 |
| 0 | 1 | – | 1 |
| 1 | 0 | 0 | 0 |
| 1 | 0 | 1 | 1 |
| 1 | 1 | – | 0 |
| 1 | 1 | – | 0 |

Debounce and synchronize clock and press.

$$q_0^+ = q'(clock' \times press + clock)$$
$$+ q(clock' \times press)$$
$$= q'(p+c) + q(c'p) = q'p + q'c + qc'p$$
$$= (q' + qc')p + q'c$$
$$= (q' + c')p + q'c = c'p + q'c + q'p$$

*(q'p is a consensus term page 90 and Theorem 8 page 47)*

**Figure E1002**

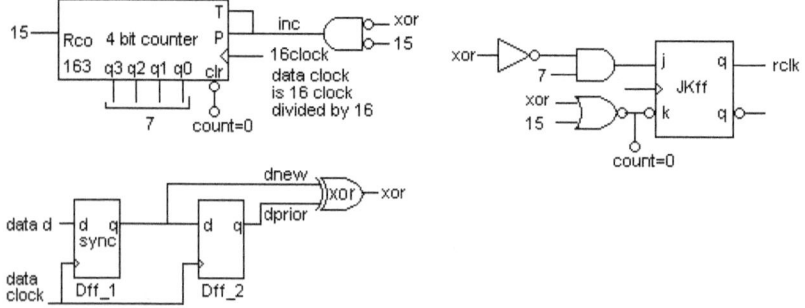

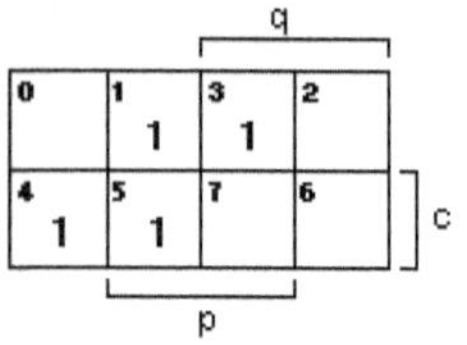

## Experiment 11

| xor | 15 | 7 | count=0 | rclk=H | rclk=L | inc |
|-----|----|----|---------|--------|--------|-----|
| 0 | 0 | 0 | | | | 1 |
| 0 | 0 | 1 | | 1 | | 1 |
| 0 | 1 | – | 1 | | 1 | |
| 1 | – | – | 1 | | 1 | |

$xor = dprior \oplus dnew$

$(count = 0) = xor'15 + xor = 15 + xor$

$j = (rclk = H) = xor' 15' 7 = xor' 7$

$k = (rclk = L) = xor' 15 + xor = 15 + xor$

$inc = xor' 15' (7' + 7) = xor' 15'$

**Figure E1103** Debounce and synchronize data d.

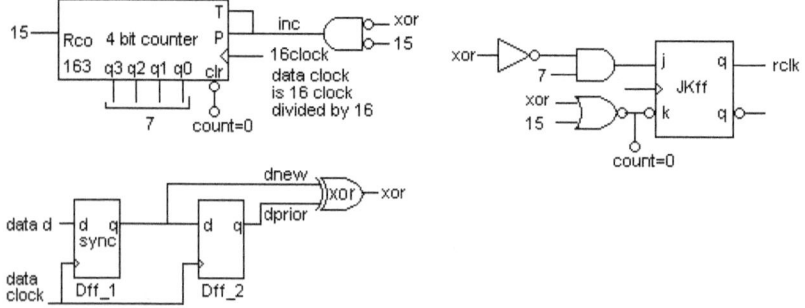

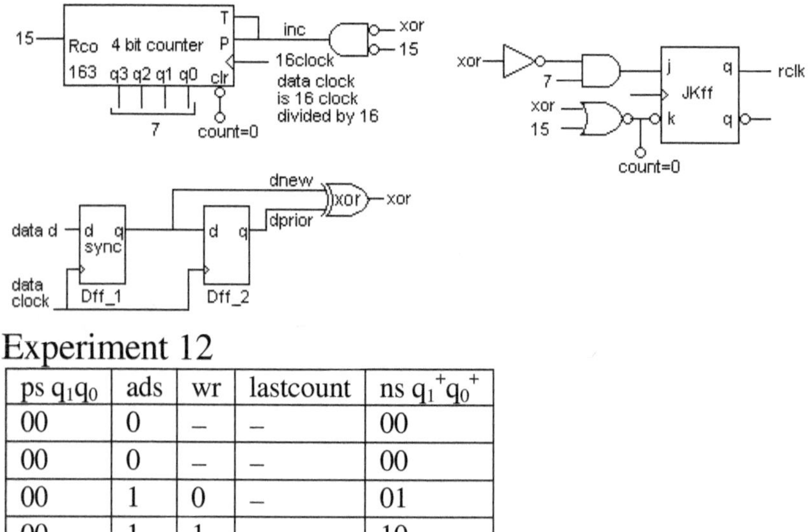

## Experiment 12

| ps $q_1 q_0$ | ads | wr | lastcount | ns $q_1^+ q_0^+$ |
|--------------|-----|-----|-----------|------------------|
| 00 | 0 | – | – | 00 |
| 00 | 0 | – | – | 00 |
| 00 | 1 | 0 | – | 01 |
| 00 | 1 | 1 | – | 10 |
| 01 | – | – | 0 | 01 |
| 01 | – | – | 1 | 00 |
| 10 | – | – | 0 | 10 |
| 10 | – | – | 1 | 00 |

$$d_1 = ads \cdot wr \cdot S_0 + lastcount' \cdot S_2$$
$$d_0 = ads \cdot wr' \cdot S_0 + lastcount' \cdot S_1$$

**Figure 1201**

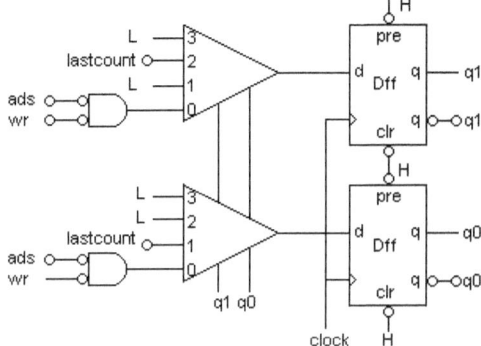

# Experiment 13

| $q_2q_1q_0$ | ads/nb | wr | $q_2^+q_1^+q_0^+$ |
|---|---|---|---|
| 000 | 0 | – | 100 |
| 000 | 1 | 0 | 001 |
| 000 | 1 | 1 | 010 |
| | | | |
| | cdone | match | |
| 001 | 0 | – | 001 |
| 001 | 1 | 0 | 011 |
| 001 | 1 | 1 | 000 |
| 010 | 0 | – | 010 |
| 010 | 1 | 0 | 000 |
| 010 | 1 | 1 | 000 |
| | | | |
| | ddone | | |
| 011 | 0 | | 011 |
| 011 | 1 | | 000 |
| | | | |
| | rfdone | | |
| 100 | 0 | | 100 |
| 100 | 1 | | 000 |

# Digital Design

$$d_2 = ads/nb' \cdot S_0 + rfdone' \cdot S_4$$

$$d_1 = ads/nb \cdot wr \cdot S_0 + (cdone \cdot match') \cdot S_1 + cdone' \cdot S_2 + ddone' \cdot S_3$$

$$d_0 = ads/nb \cdot wr' \cdot S_0 + (cdone' + cdone \cdot match') \cdot S_1 + ddone' \cdot S_3$$

**Figure 1301**

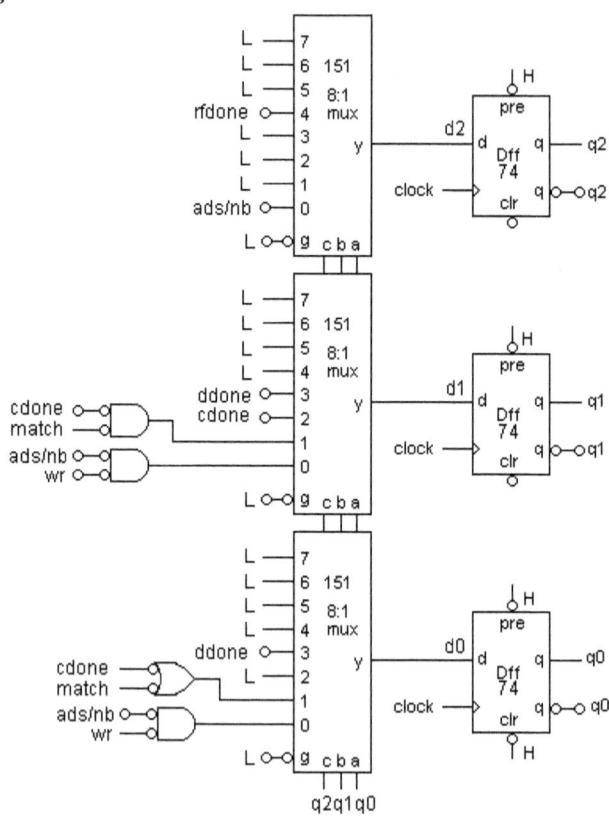

# Experiment 14

$$c_7 = k_4 \quad c_6 = k_3 \quad c_5 = k_2 \quad c_4 = k_4 + k_3 + k_2$$

$$c_3 = k_1 \quad c_2 = k_4 + k_3 + k_1 \quad c_1 = k_4 + k_2 + k_1$$

$$s_2 = c_7 + c_6 + c_5 + c_4 \quad\quad s_1 = c_7 + c_6 + c_2 + c_1$$

$$s_0 = c_7 + c_5 + c_3 + c_1$$

use trees of XOR

Two examples of syndrome calculations with 0 and 1 errors show that the *hardware interpretations* are as follows.

*Example* : $R = 1001100 + 0000000$  (*no errors*)

$$S = H \times R^T = \begin{bmatrix} 1111000 \\ 1100110 \\ 1010101 \end{bmatrix} \times \begin{bmatrix} 1 \\ 0 \\ 0 \\ 1 \\ 1 \\ 0 \\ 0 \end{bmatrix} = \begin{bmatrix} 0 \\ 0 \\ 0 \end{bmatrix}$$

*interpret S as* 000, *no errors*

*Example* : $R = 1001100 + 0000010$ (*one error*)

$$S = H \times R^T = \begin{bmatrix} 1111000 \\ 1100110 \\ 1010101 \end{bmatrix} \times \begin{bmatrix} 1 \\ 0 \\ 0 \\ 1 \\ 1 \\ 1 \\ 0 \end{bmatrix} = \begin{bmatrix} 0 \\ 1 \\ 0 \end{bmatrix}$$

*error in bit position* 2

## Safety issue - The AC Line Voltage is extremely dangerous

The voltage at an AC outlet is 120V
The USA line voltage frequency is 60 Hz. In the world it is 50Hz.
The 120VAC peak to peak voltage is 340 volts (from 1 to −1).

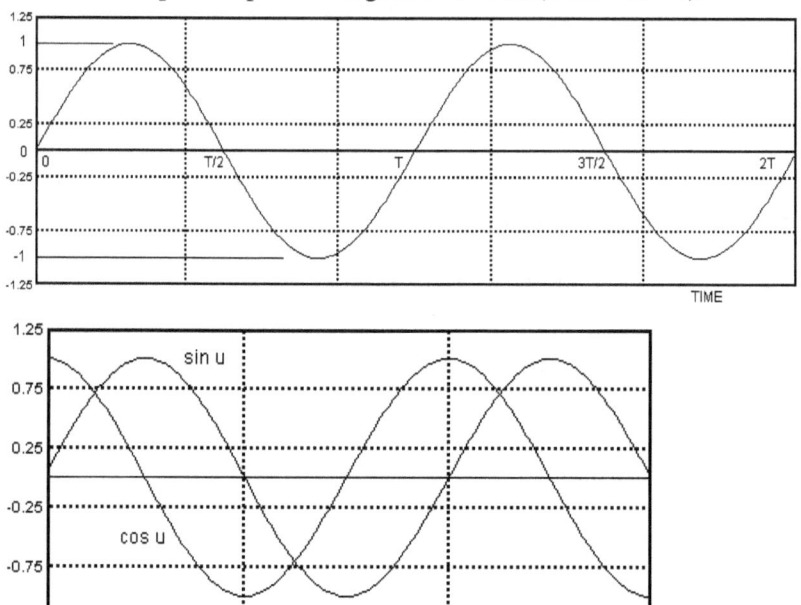

RMS Voltage

The RMS voltage for a sinusoid is that value which will produce the same power as an equivalent DC voltage.

$$Power = I_{dc}^2 R = I_{rms}^2 R \quad where \quad I_{rms}^2 = \frac{1}{T}\int_0^T i^2 dt \; average \; over \; one \; cycle$$

$$if \quad i = I_m \sin \omega t \quad then \; i^2 = I_m^2 (\sin \omega t)^2 = \frac{I_m^2}{2}[1 - \cos 2\omega t]$$

$$I_{rms}^2 = \frac{1}{T}\int_0^T i^2 dt = \frac{I_m^2}{2T}\int_0^T [1 - \cos 2\omega t]dt = \frac{I_m^2}{2T}\left[t - \frac{1}{2\omega}\sin 2\omega t\right]_0^T = \frac{I_m^2}{2}$$

$$I_{rms} = \frac{I_m}{\sqrt{2}} \quad and \; by \; the \; same \; process \quad V_{rms} = \frac{V_m}{\sqrt{2}}$$

# Index

www.ingramcontent.com/pod-product-compliance
Lightning Source LLC
Chambersburg PA
CBHW051626170526
45167CB00001B/79